OCEAN LIFE

Martin and Heather Angel

Octopus Books

Contents

The ocean environment

Over seven-tenths of the Earth's surface is covered by ocean, to an average depth of over two miles. Within this vast volume of water lives the majority of the Earth's inhabitants. The oceans affect all our lives via the weather. The immense tracts of open water are the source of the water vapour in the atmosphere that produces rain. The oceans redistribute heat, transporting it from the tropics in great currents like the Atlantic Gulf Stream and the Pacific Kuroshio into high, cooler latitudes.

These currents are massive flows of water driven partly by the wind's frictional drag on the sea surface, partly by forces produced by the rotation of the Earth and partly by density differences. In the Northern Hemisphere, the currents form great clockwise gyrals, whereas in the Southern hemisphere they form anticlockwise gyrals. The density differences are due to cold water being heavier than warm water and to variations in saltiness. At certain latitudes, the surface water becomes saltier, and so heavier, because more water is lost by evaporation than rain falls on the sea surface. At other latitudes the sea water may be diluted by heavy rains or by melting ice. At the edge of the Antarctic ice, for example, the sea water is not only cold but it is also diluted by melting snow and ice. This water sinks and can be followed northwards beyond the equator at depths of about 1000 m (3250 feet). These so-called 'water masses' have characteristic properties of temperature and salinity, as well as their own specific fauna. Thus the great current gyrals in the North Atlantic and the North Pacific contain characteristic species of marine animals, in the same way as the continents of America and Australia have their own specific land animals. Where two current systems flow side by side, it is sometimes possible to distinguish one from the other solely by the colour of the two separate water masses.

The deepest ocean water is formed near the poles by very cold surface water sinking to the bottom. This water is rich in dissolved oxygen and therefore capable of supporting life at the bottom of the deepest ocean. Early oceanographers, noting the decrease in the numbers of animals with depth, erroneously concluded that life would be totally absent below

Sea water contains salt in the proportion of about 35 to one thousand. The exact amount varies according to how much water evaporates from the sea surface through the sun's heat, and how much rain falls in a particular area. Thus by studying the temperature and the salt content of the sea water oceanographers can trace the current patterns in the oceans. The oceans both affect and are affected by the weather. Wind, sun and rain change the properties of the sea water and drive the ocean currents. The ocean currents can either warm or cool local climates.

1000 m (3250 feet). They thought the great hydrostatic pressure would cause the disappearance of animals; each 10 m (30 foot) increase in depth increases the hydrostatic pressure by one atmosphere. However, there is now ample proof of life at great depths by direct observations from submersibles, by free-fall cameras and from deep trawl samples – even from the bottom of one of the deepest ocean trenches at 10,910 m (35,800 feet) in the Challenger deep of the Marianna Trench off the Philippines.

In the tropics the sea is stratified; the warm surface water lies in a layer over the cold, deeper water. The density differences between the warm surface and cold, deep water are sufficiently large to prevent them mixing. Thus, if the surface water becomes depleted of mineral nutrients, such as nitrates and phosphates, they are not replenished from the deep water. Similarly, if the deep water has its dissolved oxygen content reduced, the oxygen is replaced very slowly by diffusion from the surface. In a few areas, such as off Peru and California, the pattern of currents results in cold, deep water welling up to the surface. This deep water is rich in nutrients and stimulates a rapid growth of the plant plankton, known as a bloom, which is grazed on by millions of animals. These areas are the centres of the world's great fisheries. Off the coast of Peru, over ten million tons of anchovies, to name one species, have been caught in a single year.

In temperate seas, the sea is stratified in summer but it becomes mixed up by the winter storms. During winter, the day length is too short to allow plant growth, so each spring there is a great bloom of plant plankton as the days lengthen and the sea warms up. The bloom dies away as the animals graze the plants; and when stratification sets up, the nutrients become exhausted.

Sea water has another property which is important in the understanding of ocean life. When seen in a drinking glass, water looks completely transparent. Pure sea water, however, absorbs light; if the water contains impurities or suspended particles, the light is absorbed more quickly. Even in clear water, all red light is absorbed at a depth of 30 m (100 feet). 100 m (300 feet) down, only one-hundredth of the surface intensity is left; this is the lowest limit at which photosynthesis occurs fast enough for plants to grow. Photosynthesis is the process by which plants use the sun's energy to combine the simple molecules of carbon dioxide and water into sugars and it is the source of all energy for life processes. 1000 m (3250 feet) down is the realm of darkness, even the blue-green light which penetrates deepest has become absorbed. At such a depth, the only light present is produced by the animals themselves.

High up above the northwestern Atlantic, astronauts look southwards down on Florida. In the far distance and under the clouds lies Cuba. On the left are the Bahamas, with the large island of Andros showing up clearly. Northwards through the Florida Straits flows the Gulf Stream. It flows rapidly up the coast before swinging to the northeast, carrying warm water across to Western Europe and ameliorating its climate. Deep beneath the Gulf Stream runs a counter current flowing to the south into the Gulf of Mexico.

Plankton – the 'grass of the sea'

Plankton is the name given to the tiny – but not always microscopic – animals and plants which drift in the sea. The plants, or the phytoplankton, are however nearly all microscopic and therefore most individuals are invisible to the naked eye. When a phytoplankton bloom occurs, the sheer numbers of minute cells can colour the sea. Only in shallow coastal waters do seaweeds, like kelp and wracks, make an important contribution to the food of herbivorous animals. In the open ocean, the microscopic phytoplankton is the equivalent of the grass in fields and pastures on land and is indeed sometimes referred to as the 'grass of the sea'. All ocean life feeds either directly or indirectly on phytoplankton, which forms the basis of all food chains in the sea.

It has been estimated that each year two million million (a thousand US billion) tons of phytoplankton are produced in the world's oceans. The animal herbivores must eat virtually all this immense production of plant material, otherwise it would accumulate on the seabed. The calcareous (chalky) and

siliceous skelteons of these tiny plants do accumulate on the seabed but only at a rate of about 1 mm every 1000 years. Even so, over millions of years, the accumulated skeletons form thick geological deposits. Chalk, for example, is formed from the miniscule calcareous skeletons of coccolithophores.

In the coastal and upwelling areas, single celled phytoplankton forms, like diatoms and dinoflagellates, are an important source of plant food for herbivorous animals. Diatoms range in size from 10μm – 1mm (1/1000 – 1/25 inch). They have glass-like skeletons made of silica, which consist of two interlocking parts. Reproduction is by simple fission – each cell divides into two. The two parts of the skeleton separate and inside each a new inner part is formed. The average size of the cells gets progressively smaller, until a form of sexual reproduction occurs, which results in the growth of full-sized cells.

Skeletons of diatoms accumulate on the ocean floor to form diatomaceous ooze. Where an old seabed covered by this ooze has become raised above the sea, the deposits are known as diatomaceous earth and are used commercially in polishes and some toothpastes. Because diatomaceous earth is inert, it is used to make highly explosive nitro-glycerine safe to handle in the form of dynamite.

Dinoflagellates have cell walls made of cellulose. The cellulose walls are sometimes arranged in plate-like mosaics which give many of the species an armoured appearance. Each dinoflagellate has a tipped flagellum and an equatorial flagellum which lies in a groove around the centre of the organism. These flagella are microscopic, whip-like structures which beat in such a way that the organism corkscrews through the water. Dinoflagellates also reproduce by means of binary, or dual, fission. Some species, including *Gymnodinium* and *Gonyaulax*, produce toxins. When these species multiply too rapidly they produce toxic 'red tides', which turn the sea red.

Toxic blooms can cause mass mortalities of fish and birds, especially in upwelling areas such as off California and south-

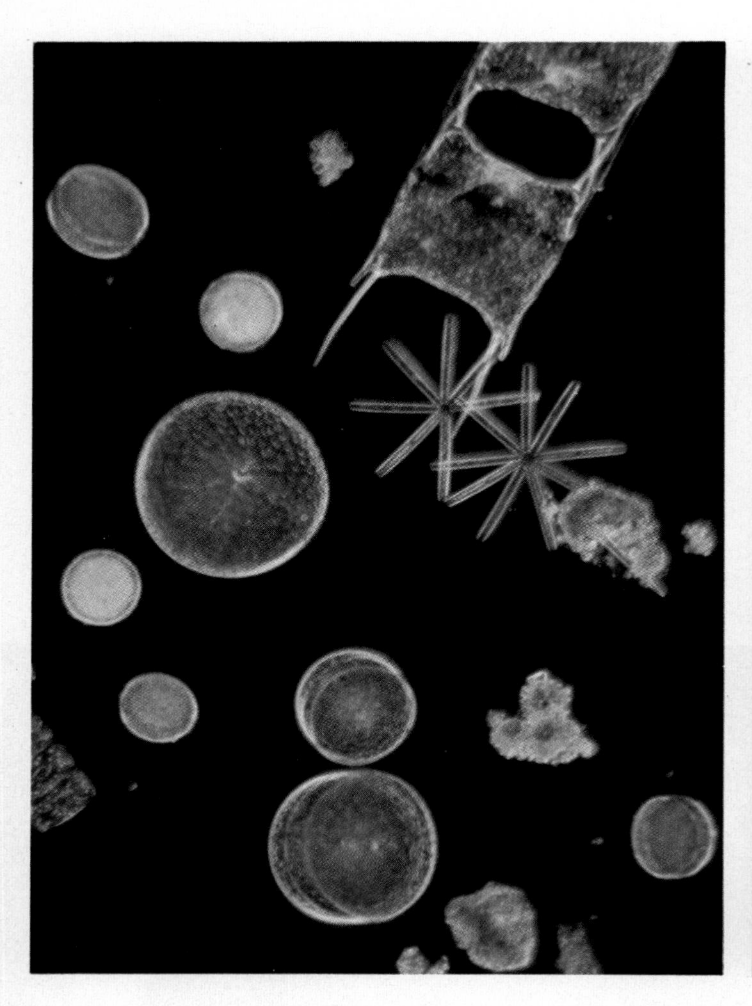

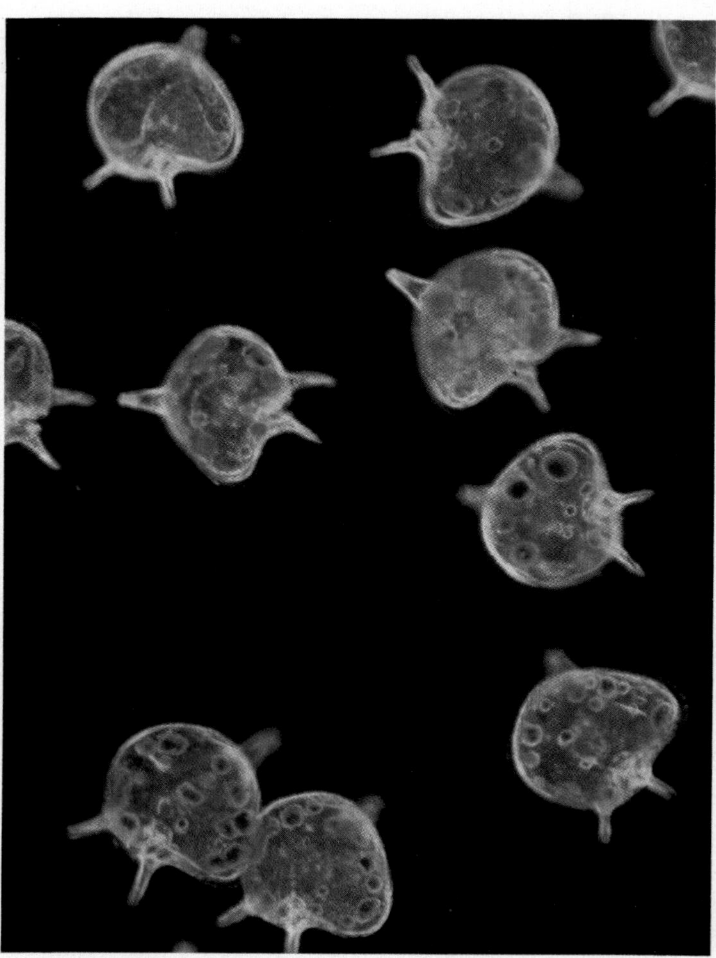

west Africa. A red tide off Florida in 1947 killed an estimated 100,000 tons of fish. In 1972, a red tide contaminated nearly 100 miles of the southeast coast of Florida. Even in the Book of Exodus mention is made of a red tide. One of the seven plagues of Egypt, which turned the Nile to blood, was a red tide which caused a mass mortality of frogs. Each year, several human deaths are caused by people eating shellfish which have been contaminated by these toxic blooms.

Many species of dinoflagellates are capable of producing brief flashes of their own light. At night, bright pinpoints of light, emitted from the surface, are visible even in coastal waters. Some dinoflagellates live inside corals and sea anemones. This association, and its mutual benefit to both the dinoflagellate and the coral, will be described in more detail in the chapter on coral reefs.

Records of fossil diatoms date back some 180 million years, to the Triassic Period. Dinoflagellate records, on the other hand, have only been traced back to the Cretaceous Period – a mere 70 million years ago! In the open ocean, tiny flagellates (organisms bearing flagella) and coccolithophores are the most important source of animal food.

The animal plankton, or zooplankton, consists of herbivores which feed directly on the phytoplankton, and of omnivores and carnivores. The herbivorous animals either have to be small themselves, as in the case of protozoans and many larval forms; or else they have to have a system of fine filters. These filters may be mucus sheets as in the pteropods (sea butterflies) and salps (transparent ocean life); or else they are made up of the fine setae on the limbs of herbivorous crustaceans such as euphausiids. One well known euphausiid is *Euphausia superba*, the krill eaten by baleen whales in the Antarctic.

Many planktonic animals are omnivores – eating anything and everything they can. *Calanus finmarchicus* is a copepod which can either filter feed by using the fine bristles or setae on the limbs near its mouth, or can seize individual particles and small prey. The carnivores in contrast, are usually easily recognized by their conspicuous jaws. The arrow worms are good examples.

Zooplankton occurs at all depths in the ocean. Herbivorous forms, however, have to live in or near the surface 100 m (300 feet) where their food, the phytoplankton, is present. Deeper down, the zooplankton inhabitants are either carnivorous or else detritus feeders eating the rain of debris and excrement which descend from the surface layers. The number of the zooplankton individuals progressively decreases as the depth increases, until below 2000 m (6500 feet) they are very sparse indeed. Many of the animals are permanent members of the plankton; but some are the immature larval forms of bottom-living animals or of the larger, fast swimming fish and shrimps.

One of the greatest unsolved mysteries about zooplankton is the daily upward migration made each evening by many species, only to descend again at dawn. Some species of ostracods, only 1–2 mm long, migrate up nearly 400 m (1300 feet) and down again each day. A remarkable feat for such tiny animals.

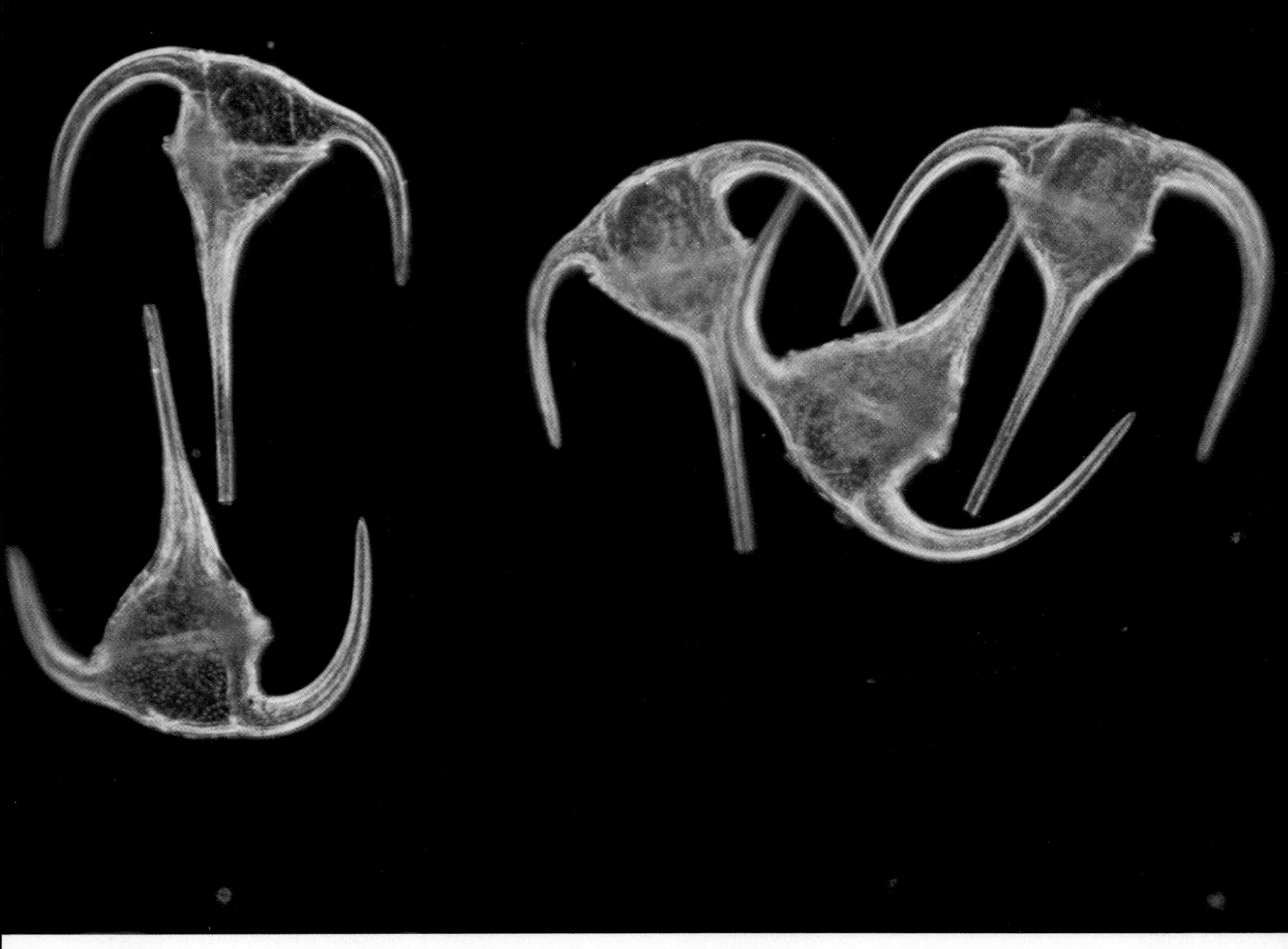

PREVIOUS PAGE LEFT
Phytoplankton is the source of nearly all food in the deep ocean. By combining the simple molecules of carbon dioxide and water together in the presence of sunlight in a process called photosynthesis, they form simple sugar molecules. These simple molecules are the building blocks and energy source for all other biological chemicals and processes. Here is a selection of diatoms from the English Channel. *Lauderia borealis* has small cylindrical cells in chains. *Rhizosolenia* has needle shaped cells. In *Thalassiothrix* the cells are rods arranged in zig-zags or stars. *Biddulphia* has rectangular cells with two pairs of projections. *Stephanopyxis* is the chain of large oval cells.

PREVIOUS PAGE RIGHT
A plankton haul taken close inshore will always contain a large variety of different crustaceans. The two largest animals are copepods which are nearly two millimetres long. The animal with the large eyes and the long spine on its back is a crab larva. The smaller animals are mostly small species of copepods and their naupliar larvae. In the centre of the picture is a cyprid larva of a barnacle. All these animals are herbivores, sieving the microscopic cells of phytoplankton from the water. Notice how most of these planktonic animals are transparent, only their eyes and gut contents showing much colour.

FAR LEFT
Each year in temperate and polar regions, after winter storms have churned up the water, there is a big outburst of phytoplankton. This is a magnified picture of some diatoms which form a typical bloom in the English Channel. The two star shaped colonies of *Thalassiothrix* lie next to two *Biddulphia* cells. The round pill-box cells are of *Coscinodiscus*. Their cell walls are silica-like and in areas of the deep ocean, masses of their skeletons form a bottom sediment of diatomaceous ooze. Terrestrial deposits of this ooze form a mineral, kieslguhr, which is used in the manufacture of dynamite. The largest cell of *Coscinodiscus* in the picture is about a fifth of a millimetre in diameter.

LEFT
Peridinium depressum is a dinoflagellate. The cell wall is made up of plates of cellulose fitted together into a mosaic. It has two flagella, one that runs transversely round its middle in a groove called the girdle, the other projecting posteriorly, close to the posterior pair of spines. Each of these cells is about a tenth of a millimetre across. In common with many dinoflagellates, they produce brilliant flashes of luminescence. They swim quite actively and migrate up and down a few metres each day. Most dinoflagellates are armed with stinging organelles called trichocysts, and indeed some species are carnivorous.

ABOVE
Ceratium tripos is a very common dinoflagellate that typically occurs in temperate waters. It is particularly common in late summer and early autumn. As in most dinoflagellates, it multiplies by simple fission, splitting in two. The recently divided cells sometimes stick together forming long chains. The warmer the water, the longer the spines tend to grow. At certain times of year they form resting cysts, and these cysts are now being studied in geological deposits. Some dinoflagellates live as symbionts in corals. Other species are highly toxic and cause red-tide mortalities.

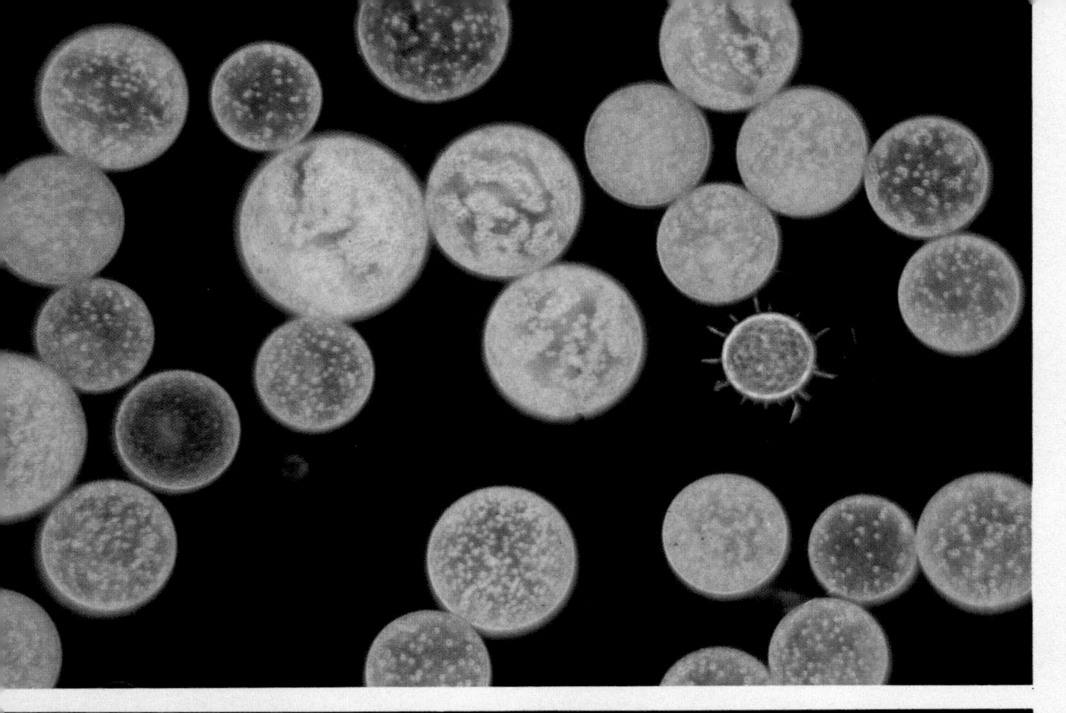

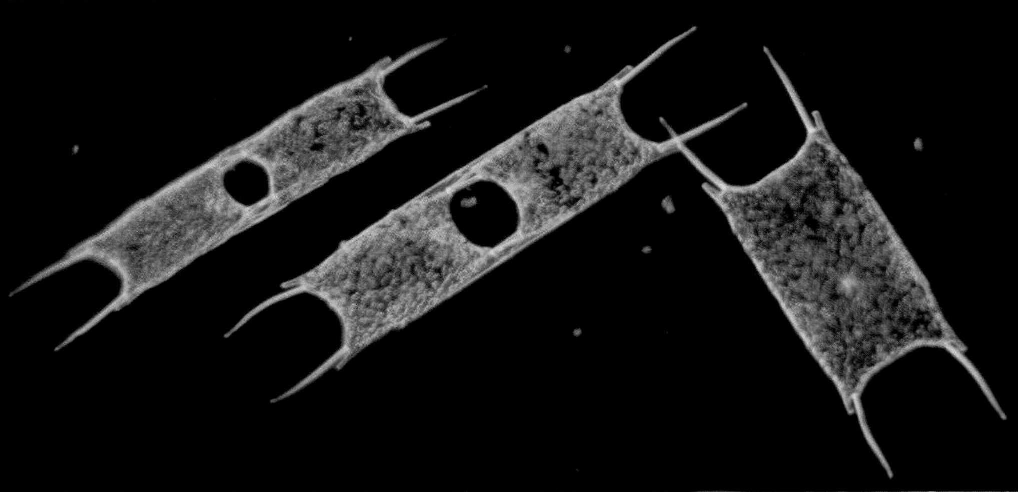

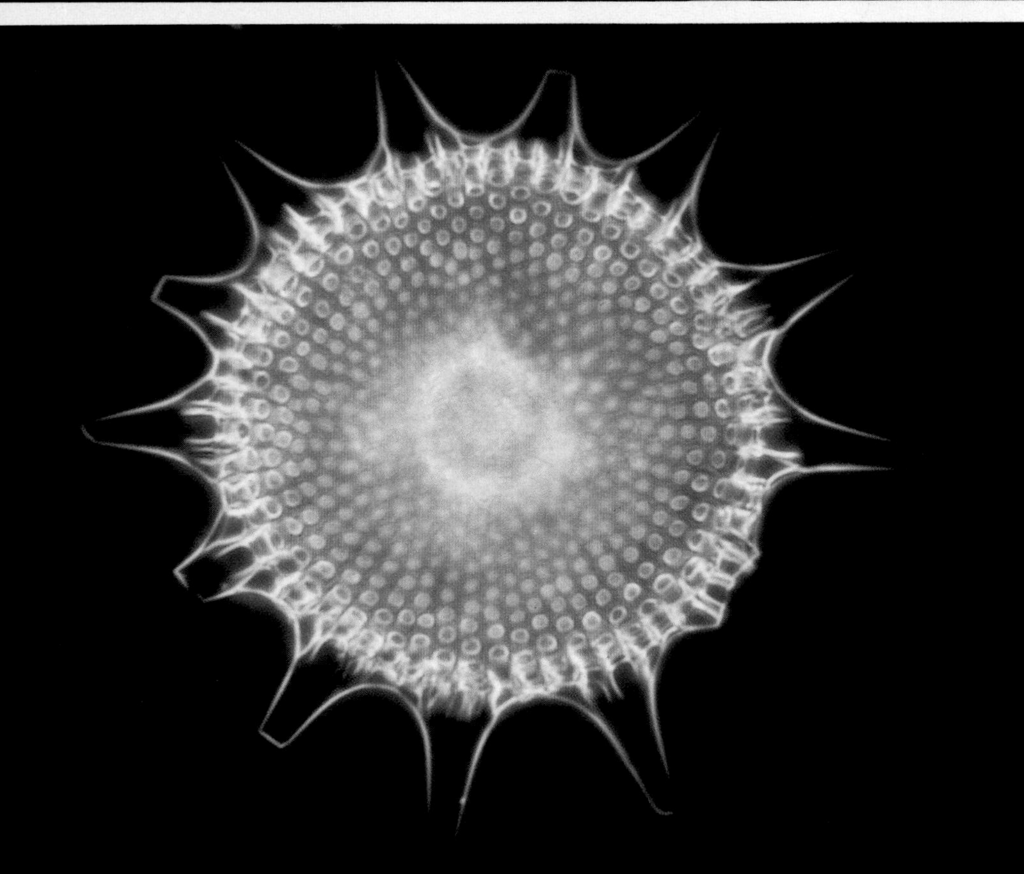

LEFT
The brilliant green spheres of *Halosphaera viridis* are yellow-green algae, and can grow to nearly a millimetre in diameter. They multiply by dividing into masses of tiny spores each carrying a fine whip-like flagellum. They are often caught in nets very deep down, and it has been suggested that they mark the water that has spilled out from the Mediterranean through the Straits of Gibraltar into the Atlantic. However, more normally they occur right at the very surface and when present in large numbers give the water an oily appearance.

CENTRE
The large cell of *Biddulphia sinensis*, and the two pairs of cells still linked together after division, are diatoms. This was a well known species in the Indo-Pacific region but was unknown off Europe until 1903, when it was found in the Heligoland Bight. By 1909 it had spread throughout the North Sea and now is one of the commonest diatoms there. The shells or frustules of such diatoms are ornamented with very fine striations and punctures that are used to test the resolving power of light microscopes. Electron-microscope studies show that the shell has a geodesic dome rib structure. Engineers who design bridges use similar structures to minimize stresses.

BELOW
Radiolarians are marine protozoa that are related to *Amoeba;* there are many planktonic species that are sometimes so abundant that the bottom sediments are composed entirely of their skeletons. Here the delicate sculpturing of the silica-like shell is shown. When living, there may be several nuclei inside the shell, and radiating out from the pores are fine pseudopodia. These pseudopodia are fine processes of protoplasm which engulf the animal's food particles. They may extend out two or three times the shell's diameter. Many species contain within them symbiotic algae. Surprisingly little is known about the biology of these elegant animals.

RIGHT
Foraminifera are marine amoeboid protozoa with shells of calcium carbonate. Over large tracts of the deep ocean the bottom sediments consist mainly of these shells, in what is called *Globigerina* ooze. Palaeontologists study these shells in core samples of marine deposits. The foraminiferan remains are the fossils of older and older plankton communities the deeper the core penetrates. By comparing their modern distributions with present water conditions, the water conditions prevailing when the deposits were laid down can be deduced. These foraminiferan shells form a sample of *Globigerina* ooze and are about a millimetre across.

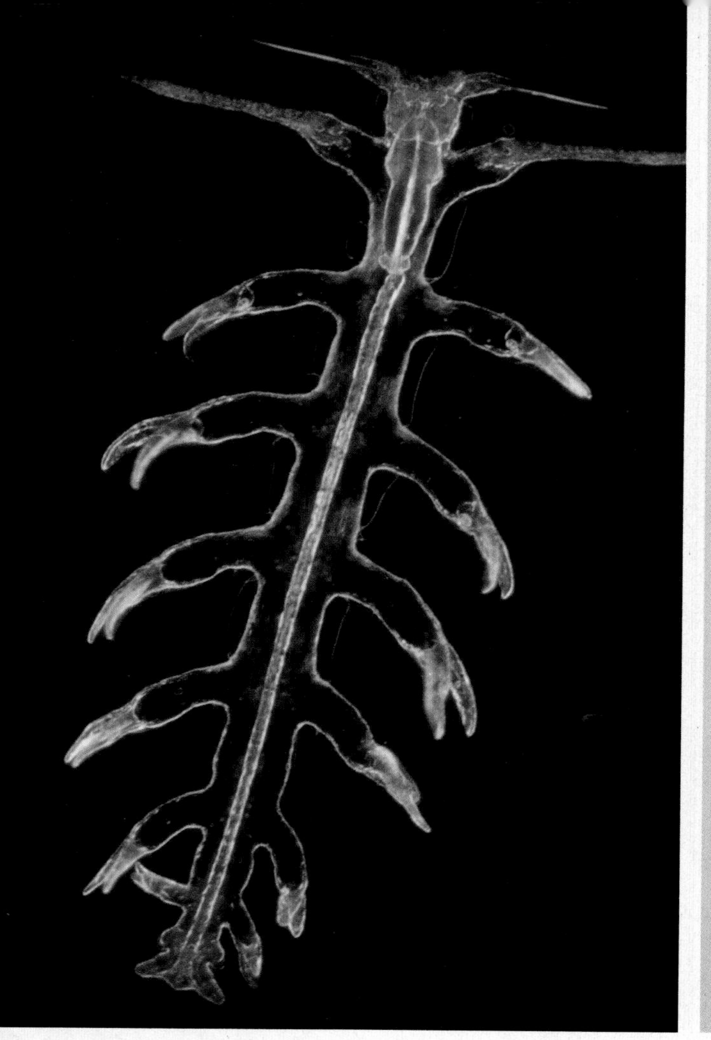

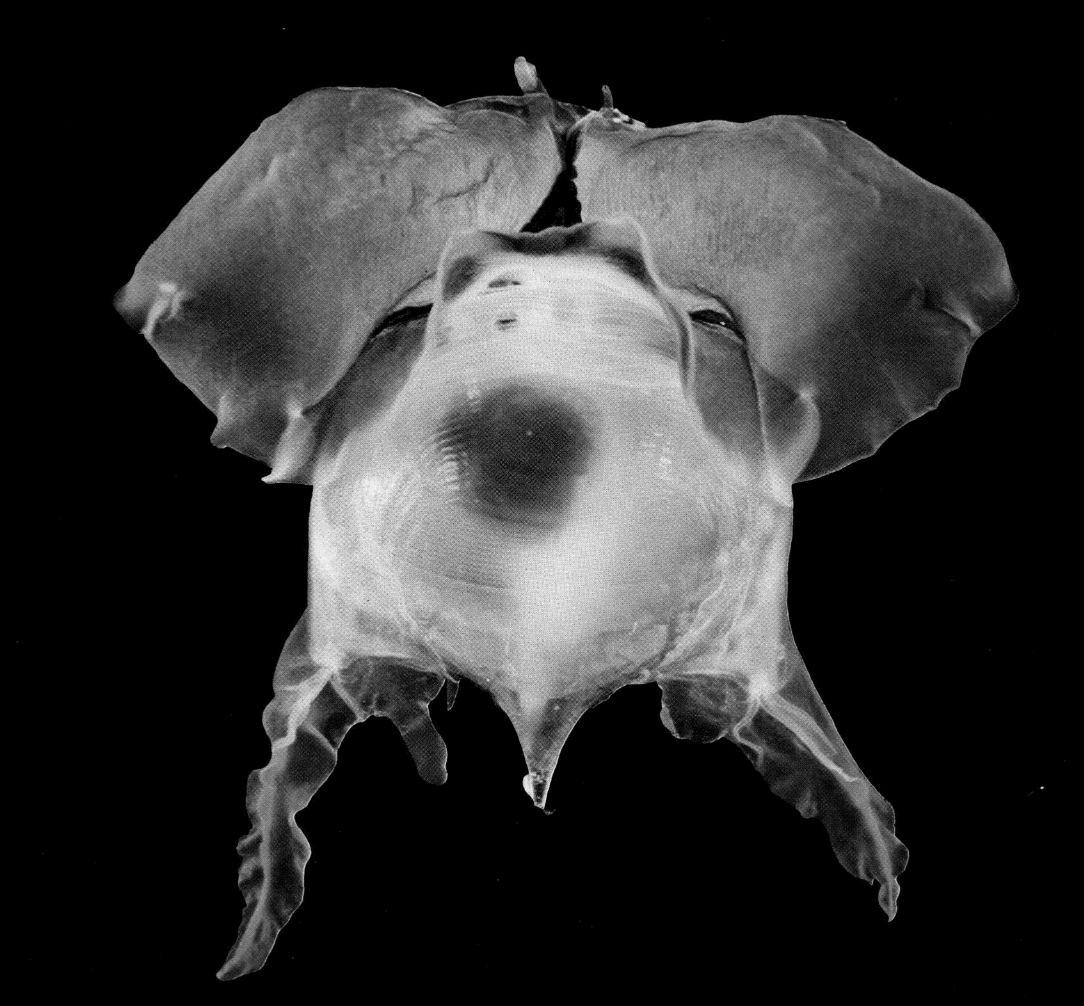

TOP LEFT
This young specimen of *Tomopteris helgolandicus* is a planktonic polychaete worm. The body is segmented, each segment having a flattened lobe or parapodium on either side. The animal swims by means of rapid snake-like waves which pass rapidly up the body from tail to head. These delicate transparent animals are carnivores, eating small copepods and other planktonic organisms. The gut of this particular specimen is quite empty. Occasionally specimens are found with a little parasitic medusa swimming round in its blood system.

TOP RIGHT
Planktonic ostracods are little bivalved crustaceans which are extremely abundant in oceanic waters, but often completely absent from shallow coastal areas. Most of the species are less than three millimetres long, but this large red species is twice that length. *Conchoecia valdiviae* is a carnivore eating other crustaceans and occurs at depths of 600–900 m (1950–2950 feet) in the equatorial waters of the Atlantic. It swims rapidly using two antennae that project from the notch in the anterior end of the shell. Some of these tiny organisms migrate 400 m (1300 feet) up towards the surface at dusk, and down again at dawn.

LEFT
Cavolinia tridentata is a pteropod, a type of planktonic snail. It swims by beating its flap-like wings, which inspire the colloquial name of sea butterfly for these little molluscs. The yellow plumes projecting from the posterior of the shell are retractable. They help to slow the rate at which the animal sinks by acting as draglines. The wings are covered by fields of minute hairs or cilia, which move sticky sheets of mucus over the wings and into the mouth. The mucus traps detritus and phytoplankton which are the pteropod's food. In rich upwelling areas such as off the north-west coast of Africa, the bottom deposits are made up of millions of pteropod shells.

ABOVE
Phronima is a transparent little amphipod, related to sand-hoppers, which is about 2 cm long and builds itself a barrel out of the swimming bell of a siphonophore or the test of a salp. The barrel is open at both ends and the *Phronima* drives a current of water through the barrel so that it is propelled along. The barrel is used as a nursery, and here several hundred juveniles are massed in a crescent round the adult as we look into the front opening of the barrel.

BELOW
This siphonophore *Physophora hydrostatica* is abundant in the Sargasso Sea. It is a colonial form with the individual polyps modified for special functions. At the top of the stem or rachis is the gas-filled float, below which are rows of swimming bells. The pink tentacles are armed with batteries of stinging cells and they vigorously sweep the water if any disturbance is felt. Below the tentacles are the feeding and reproductive individuals. These and related siphonophores have been observed concentrated at depths where deep-scattering layers are recorded. These are zones where echo-sounders pick up heavy reflections of their sound transmissions, presumably reflected by the gas floats.

BELOW RIGHT
Within the plankton some of the most beautiful and strangely shaped animals are larval forms. This is the pluteus larva of the heart urchin *Echinocardium cordatum*. The long arms, which are stiffened with calcareous rods, are covered with minute cilia, microscopic hair-like organs whose beating keeps the larva afloat. After spending several weeks in the plankton feeding on phytoplankton, and after being carried many miles away from where it was released, the larva sinks to the bottom and changes into the adult form. Many bottom living animals have planktonic larval stages so that they can be dispersed as widely as possible.

BOTTOM
Thysanopoda species are some of the largest euphausiids. The record size for these shrimp-like animals was a 15 cm long deep-living specimen from the Pacific. The commonest shallower-living species are 1–5 cm long when adult, and many of them are important herbivores. Perhaps the most famous of these animals is the krill, *Euphausia superba*, which is the staple diet of the rorqual whales in the Antarctic. This related species aggregates into dense swarms near the surface, discolouring the water. Even some of the temperate species form swarms that are dense enough to block the filters on the intakes to ship's engines.

RIGHT
This exquisitely coloured animal is the 2 cm long larval stage of a mid-water decapod prawn *Oplophorus*. The plankton contains many larval forms of the mid-water animals. Many of the larvae of carnivorous adults are herbivores. The change of diet that accompanies the metamorphosis into the adult form must be an extremely critical stage. This is the time when the greatest mortality occurs when attempts have been made to rear these larvae experimentally.

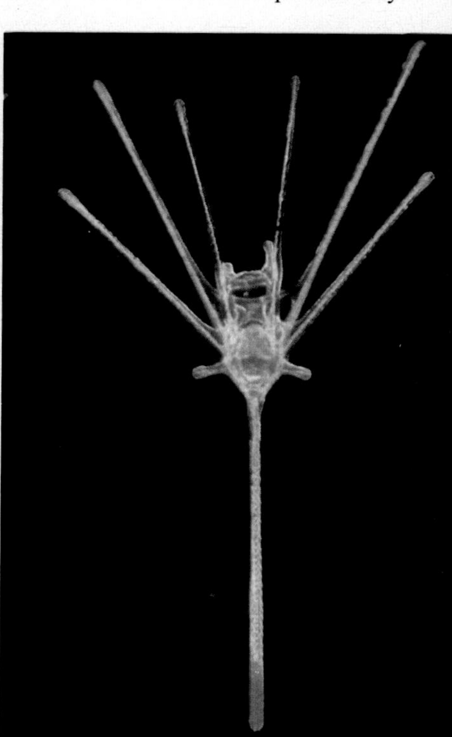

Many bottom living species have larval stages, and species that normally live on the continental shelves may occur well out in the open ocean in their larval form. These larvae are probably able to delay their metamorphosis, until they eventually find a suitable bottom on which to settle.

BELOW

The large jellyfish *Rhizostoma pulmo* has a shoal of juvenile fish accompanying it. This particular jellyfish is a filter feeder. The swollen tentacles arranged round the mouth on the underside of the bell are pierced by a fine system of canals, through which water is pumped and any fine particles extracted. Associations between fish and jellyfish are common, even if the jellyfish is carnivorous with powerfully stinging tentacles. The fish are either immune to the stings or else manage to avoid contact with the tentacles, thus gaining protection from predators.

The surface dwellers

Despite the accessibility of the sea surface, only within the last decade did the specialization of life in this zone become fully appreciated. Because early samples of surface life were collected by towing nets in a ship's wake, they were heavily contaminated with animals from deeper water, thereby masking the striking differences between them and the surface living animals. Furthermore, at night there is a massive influx of migrants into the surface layer from deeper down. Now, by towing skid nets from booms mounted on the ship's bow, undisturbed water can be sampled from the top 10 cm (4 inches). The collective term neuston is used by biologists to describe the zooplankton associated with the surface layer of ocean waters.

The surface is very rich in food. The bubbles produced by breaking waves accumulate all kinds of organic material, and then collects as a thin scum on the surface, providing a habitat for a host of bacteria and protozoa. These micro-organisms are the food of larger animals like pontellid copepods.

Pontellid copepods demonstrate the deep-blue colour which is typical of the surface fauna in the tropics and subtropics. They are curious in having two types of eye; one is adapted

for vision below water, while the other sees through the surface. Because sea birds prey on surface animals, it is of great advantage to the copepods to see above the water. They are difficult to keep in captivity since their reaction to danger is to leap out of the water and containers. Flying fish perform the same escape reaction and use their well-developed pectoral fins to glide away from danger. Even a species of squid has learnt the trick of flying as a means of escape.

Another predator of the copepods is the by-the-wind sailor *Velella. Vellella* is a colonial animal which belongs to a group related to jellyfish and sea anemones called the Siphonophora. It has a gas-filled, flattened, oval float and a triangular 'sail' that projects above the water. The underside is covered with tentacles which are armed with stinging cells or nematocysts. In the middle of the tentacles is a group of feeding individuals.

Another common siphonophore is the notorious Portuguese-man-o'war, *Physalia.* Its float can be up to 15 cm (6 inches) long and looks, from a passing ship, like a light bulb floating on the surface. The gas inside the floats of these strange animals has a high content of carbon monoxide; the gas is toxic to us because it combines with the haemoglobin in our blood. Both *Vellella* and *Physalia* have two forms, one with a sail set so that the animal tacks to the left of the wind direction and the other so it tacks to the right. Both forms are able to dip their sail and steer a zig-zag course so that they can sail to and fro across windrows in which their food tends to accumulate.

Below *Physalia's* float stream tentacles up to 10 m (30 feet) long. Since the Portuguese-man-o'war feeds on fish, its nematocysts are large and powerful. When discharged, they inject a paralyzing poison into the prey. Swimmers should be wary of brushing against the tentacles. The stings are painful and if numerous can cause temporary paralysis. Such paralysis has been known to result in drowning; children are particularly vulnerable. The most dangerous jellyfish, which uses similar stinging cells to catch its food, is the Queensland sea wasp. Although not a true neustonic animal, the sea wasp appears often near the surface. Off the coast of Australia, sea wasps have claimed 60 victims within the last 25 years, as compared with the 13 lives lost as a result of shark attacks.

Despite the powerful stinging cells of *Physalia*, a little blue and silvery fish called *Nomeus* lives in amongst the tentacles.

These fish appear to be immune to the siphonophore stings, and so they can hide safely among the tentacles without fear of attack by predators.

These stinging cells are also no deterrent to siphonophore predators; for instance, the nudibranch (or sea slug) *Glaucus* feeds on *Velella*. It is able to eat the tentacles without discharging the nematocysts. The undischarged nematocysts are enveloped by cells in the nudibranch's stomach and carried into the skin covering the lateral projections of the body. There, the siphonophore's stinging cells are used by the sea slug for its own defence. *Glaucus* solves the problem of how to stay at the surface by periodically swallowing a gulp of air.

Another molluscan predator of *Velella* is the 2 cm (one inch) long blue snail *Janthina*. This animal, which has a beautiful pale blue shell, keeps afloat by producing a 2 cm long bubble float. When seen from a ship, the *Janthina* float resembles a floating cigarette end. Occasionally, odd specimens are carried northwards in the Gulf Stream and become washed ashore on the southwest coasts of Britain.

Goose, or stalked, barnacles also drift around in the open ocean. One species, *Lepas fascicularis*, produces a bubble float like *Janthina*; but most oceanic goose barnacles keep afloat because the late larval stage (the cypris) settles on any floating object, from wood and bottles to pumice and even lumps of fuel oil. Like their well-known rocky-shore relatives the acorn barnacles, goose barnacles are crustaceans. The fertilized eggs hatch into free-swimming, naupliar larvae. After several moults, these larvae develop into cypris larvae that select a suitable substrate on which to settle down again. Under optimum conditions, a goose barnacle can grow to 2 cm (one inch) in length and start reproducing itself within three weeks of the cypris larva having settled and metamorphosed. Goose barnacles derived their name from a medieval myth which held them to be the progenitors of barnacle geese. The barnacle's feeding limbs were supposed to grow into the legs of a bird. Gradually, as it fell into the sea and developed feathers the complete body of the bird was formed!

Ocean striders (*Halobates*) are the only insects which occur on the open sea. These tropical forms closely resemble the pond skaters found on freshwater ponds and lakes. They lay their eggs on any floating objects. Despite their abundance little is known about ocean striders, including what they feed on. They will not survive submerged in sea-water.

The only large mid-ocean seaweed of importance is the yellowish-brown *Sargassum* weed. This weed normally lives attached to the bottom on tropical shores; but if broken off by storms, it survives as great floating rafts on the surface, buoyed up by its small air bladders. In the Sargasso Sea, the ancient mariners feared that they would become becalmed and permanently trapped by these *Sargassum* rafts.

Sargassum weed has its own special fauna. A total of more than 50 species have been counted, including fishes, sea anemones, hydroids, crabs, shrimps, goose barnacles and sea slugs, all of which are adapted to this special habitat. In general, the deep ocean lacks the variety of microhabitats that characterize environments on land. In a deciduous woodland there are tree-holes, rotten logs and leaf litter – to mention only a few microhabitats. Only among the surface fauna, and in particular the fauna associated with the *Sargassum* weed, is there evidence of the evolution of highly specific adaptations to a specialized environment. Take a *Sargassum* weed fish like *Histrio* away from a lump of weed and its colour and shape make it stand out like a sore thumb. Put it back with the weed and it almost vanishes.

Sargassum weed also demonstrates clearly the phenomenon of windrow formation. When a gentle wind blows over the sea surface, it induces the formation of a system of counter-rotating cylinders of water, with axes of the cylinders parallel with the wind's direction. Anything floating on the surface gets carried to where the surface water sinks, so that parallel rows of debris accumulate. This is rich feeding for the surface dwellers which also accumulate in the windrows.

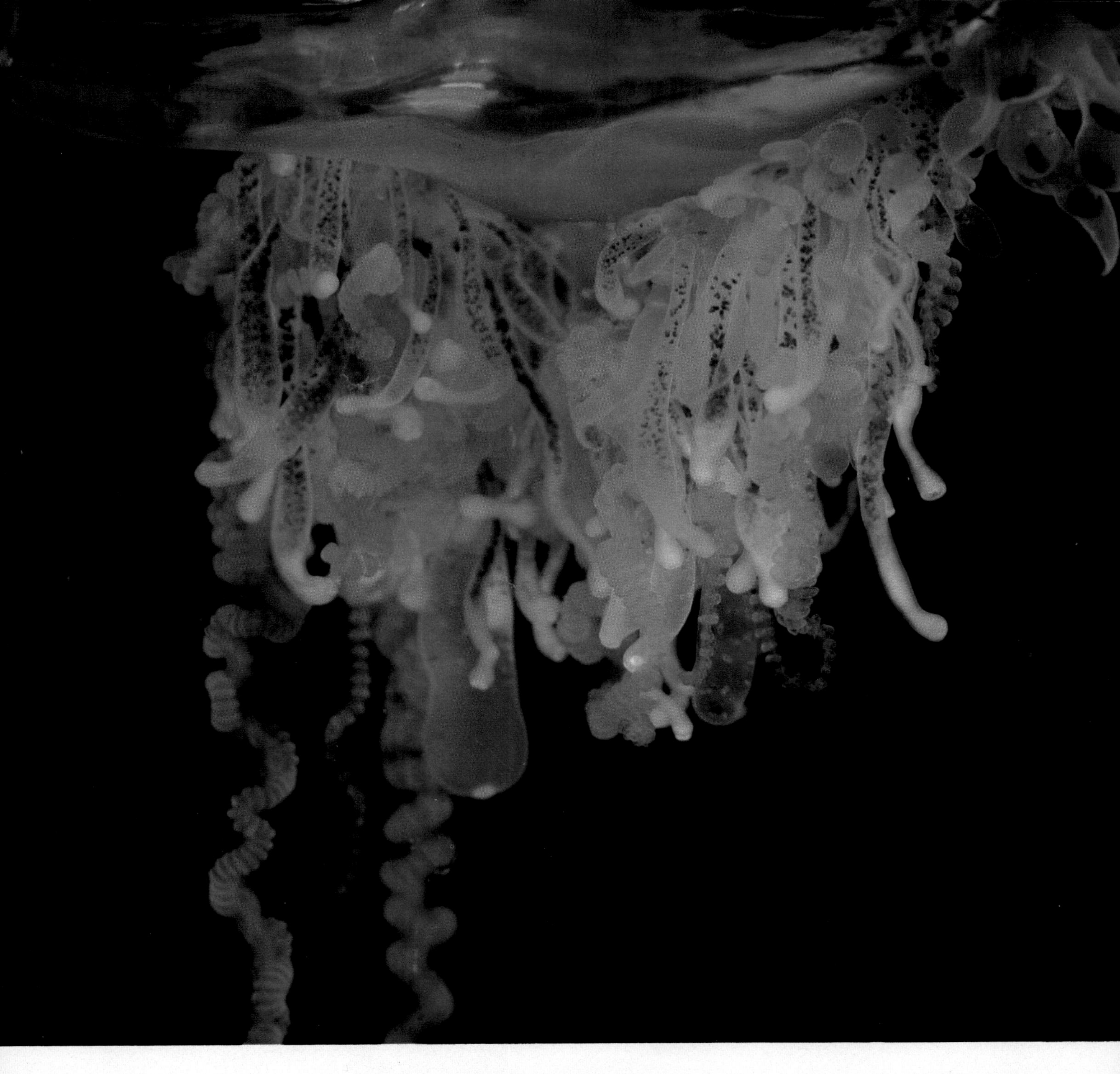

PAGE 16
Trichodesmium is a blue-green alga which often occurs in blooms that float like a reddish scum on the sea surface. It is thought that the Red Sea got its name from these reddish blooms. The alga itself is microscopic. It is filamentous and the filaments tend to be stuck together into bundles. In tropical areas like the western Indian Ocean, where this picture was taken, the alga's ability to fix nitrogen into nitrate is very important in improving the productivity of these otherwise barren areas. The lines of the algae show the windrows.

PAGE 17
This mass of blue pontellid copepods gives an impression of how vividly coloured is the plankton that lives in the top 5–10 cm (2–4 inches) of tropical and subtropical seas. The colour may be a protection against the damaging effects of the ultra-violet light in bright sunlight. Alternatively, as birds are one of the main predators of these surface-dwelling animals, blue animals viewed against the deep blue of the deep ocean will be camouflaged. These copepods are good jumpers and can leap clear of the water.

LEFT
Blowing along in the wind into St. George's Harbour, Bermuda, is a Portuguese man-o-war *Physalia physalia*. Below the float, glinting in the sun, are the long trailing dark blue tentacles. Every so often the float will bend over and dip into the sea to prevent itself from drying up. The float is set at an angle to the main axis so that as the wind blows it tacks to one side. Some specimens are left-handed and others are right-handed. Surprisingly, the gas inside the float contains a high concentration of carbon monoxide.

ABOVE
Below the float of the Portuguese man-o-war hang not only the long stinging tentacles, but a whole array of other types of polyps. Some are reproductive, others are merely protective sheaths, and others act as mouths and stomachs for the colony. When a fish is captured and killed by the stings, the tentacles holding it slowly contract until it comes within reach of the feeding polyps. The feeding polyps invest the prey and slowly digest it away. These siphonophores can be of danger to swimmers.

LEFT
Glaucus is a nudibranch snail which crawls along on the underside of the water surface. It keeps itself afloat by periodically taking a great gulp of air. It feeds on any sailor-by-the-wind or *Porpita* it encounters. Not only is it immune to the stinging cells of its prey, but it is also able to absorb them undischarged. The stinging cells are carried by blood cells into the lateral projections, or cerata, of the body, where they are used by the *Glaucus* for its own defence.

BELOW
Velella or the sailor-by-the-wind is another colonial siphonophore. It has a calcareous gas-filled float with a sail on it that projects above the sea surface. The tentacles on the underside catch its food of small crustaceans and larval fish. Despite the stinging cells on the tentacles, this *Velella* is being eaten by a violet sea snail *Janthina*. The young snail initially crawls along the surface film, but then begins to secrete a bubble float of mucus. The float both keeps the snail afloat and carries the snail's eggs.

RIGHT
Porpita is one of the most elegant and fragile of the surface living siphonophores. The chambered calcareous disc is gas-filled and is up to 2 cm (1 inch) in diameter. The tentacles, armed with stinging cells, periodically flex combing the water for prey. The feeding polyps are arranged under the centre of the disc. *Porpita* has plant cells living symbiotically inside its tissue. Perfect specimens are very difficult to obtain; any caught in nets are stripped of their tentacles. Even this specimen scooped with a bucket straight out of the sea has shed a few of its tentacles.

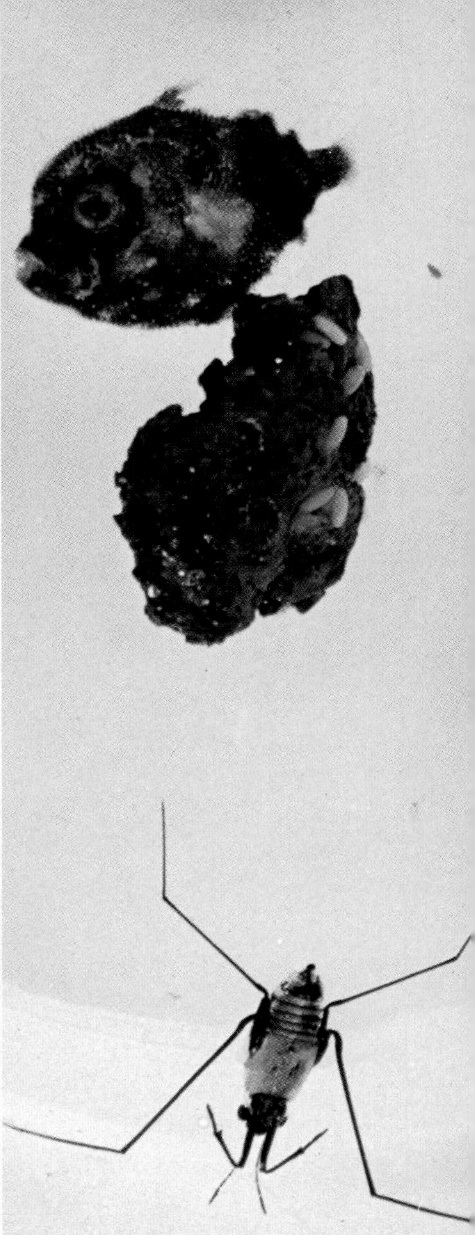

LEFT
Over wide areas of the tropical oceans, free-floating masses of *Sargassum* weed occur. It is the only large seaweed which plays an important role in the productivity of the deep ocean. Here a floating raft collects in a windrow in the Sargasso Sea. Windrows run parallel to the wind's direction and are caused by the wind inducing counter-rotating cylinders of water to be set up near the sea surface.

ABOVE
Pollution of the sea surface by fuel oil is not disadvantageous to all animals. Here a little fish larva mimics a piece of fuel oil, both in colour and by its most unusual behaviour of floating on its side at the surface. On the fuel oil are laid the eggs of the ocean strider *Halobates*. Ocean striders are the only oceanic insects. They are restricted to the tropics and subtropics. The adults are wingless and it is not known what they feed on.

ABOVE RIGHT
Beneath the water's surface the structure of

the *Sargassum* weed can be seen. The weed originates from weed beds in shallow water in the tropics. Storms break it free and, buoyed up by gas bladders, it floats out into the open ocean. The free-floating weed reproduces vegetatively and begins to accumulate a collection of specialized animals and plants. Here the weed has been colonized by an anemone and a group of hydroids, which appear as a fine fuzz on some of the leaflets.

RIGHT
Goose barnacles have planktonic larvae which eventually settle on any object floating at the sea surface. Here are a group of *Lepas fasicularis* on a piece of *Sargassum* weed. The legs are spread wide like a fishing net and occasionally flick inwards capturing small animals and pieces of debris. This particular species is capable of secreting a bubble float from the foot. Growth can be very rapid—in less than 3 weeks a newly settled larva 2 mm long can grow into a 2 cm long adult, already reproducing.

Life in mid water

All life in the sea is dependent on the plant production within 100 m (325 feet) of the surface as its ultimate source of food. Each time any food is consumed up to 90% of the energy it contains is used by the consumer and is no longer available to be passed on to the next link in the food chain. Thus the deeper the ocean gets, the less food is available to the inhabitants. Temperature also decreases with depth – apart from some polar seas where at certain times of year the whole water column is at the same temperature. Similarly, the amount of daylight penetrating decreases with depth until at about 1000 m (3250 feet) it is no longer detectable.

This reduction in light, together with the change in its colour make-up, has a marked effect on the animals' colouration. In the surface layers, fish tend to be coloured like herring and mackerel. At 250–700 m (800–2425 feet) the fish tend to be silvery-sided and deeper still they are mostly dull black or bronzy. The significance of these colour changes will be described in a later chapter.

Light is important in other ways. The red wave lengths of

the daylight spectrum are absorbed in the upper layers of the sea and only the blue-green wave lengths penetrate to the greatest depths. Hence, the animals are only sensitive to light of this colour.

At depths where light is very dim, the eyes of many deep-sea animals are large and in some fish are developed into complicated telescope-like structures. The quantity of rhodopsin, the visual pigment, is almost ten times that in the human eye. With the added advantages of wider pupils and more transparent eye fluid, these fishes are 15–30 times more sensitive to blue-green light than the human eye.

Below 1000 m (3250 feet) there is no daylight. The animals themselves produce the only light that is present. These animals tend to have much smaller eyes than the inhabitants above this zone. Closer examination of these small eyes shows that the retinas are regressed and not likely to be functional; in other words, the animals are effectively blind. Eyes, like all sense organs, require energy in order to function. With so little food available, the animals need to reduce their energy requirements to an absolute minimum. Thus these bathypelagic animals (i.e. those living below 1000 m) have evolved as compromises between the need for elaborate sensory systems to locate available food and the necessity of conserving energy.

Many fish near the surface have swim bladders, gas-filled organs which maintain the fish's neutral buoyancy, so that when a fish stops swimming, it neither sinks nor floats. They can afford heavy, calcified skeletons that act as the anchoring framework on which their powerful swimming muscles can function. They have well developed brains, livers, hearts and other organs.

As the depth increases and the food becomes scarcer, the fish cannot afford swim bladders so that heavily calcified skeletons and powerful body muscles become reduced. Below 1000 m (3250 feet) most of the fish have no swim bladder. To maintain neutral buoyancy, the skeletons are uncalcified and the body musculature reduced. Even these measures are not enough and deposits of fat are laid down or the salt content of the blood is controlled by either excluding sulphate ions or

concentrating ammonium ions. Other types of animals show parallel modifications; prawns and squid are flabby and gelatinous. Eating them raw is like consuming watery cardboard!

Gelatinous structures are usually associated with regulation of the ions in the blood. Cranchid squid, for example, are globular and gelatinous; and their blood, which contains high concentrations of ammonium ions, accounts for two-thirds of the animal's weight. The reduction of the sulphate concentration to 40% of the surrounding sea water reduces the animal's live weight by about one milligram for every millilitre of body volume. Hence, by increasing its volume and becoming gelatinous, an animal can achieve neutral buoyancy.

Another method of maintaining neutral buoyancy occurs in the cephalopod *Spirula*. This relative of squid and octopuses has within its body a spiral calcareous shell. The shell contains sealed chambers which are filled with gas at atmospheric pressure. The adults occur down to a depth of 1000 m (3250 feet). experiments have shown that the pressure at 1250 m (4000 feet) would cause the shell to implode.

In deep-living fishes the one organ that is not reduced is the mouth. Mouths tend to be huge and armed with big backwardly curved teeth. Once the prey starts to be swallowed, it cannot escape. Stomachs are enormously extensible and it is by no means unusual to find a fish containing prey equal in size to itself. At these depths it is very much a world of eat or be eaten.

However, since these deep-living animals have such poor muscle systems, they cannot indulge in long, active chases of their prey. Instead the prey is enticed slowly towards the sluggish hungry animal by a luminous lure. Some of these lures are quite elaborate and resemble animals such as copepods.

In angler fishes not only does the lure attract the prey but the female's lure attracts the tiny male. In contrast to the big lumbering female with poor eye-sight and an insatiable hunger, the male has big eyes, well-developed taste organs and powerful swimming muscles. The male's problem is to find the female before he runs out of energy and dies of starvation. He actively hunts out the female, probably locating her initially by tasting attractive chemicals she releases into the water and then by sighting the flashes of her lure. He must dodge the female's attempts to eat him and attach himself to her flank. There the male degenerates into a parasitic sex organ permanently fixed to his female mate.

Deep-sea animals are so sparsely distributed that the problem of any one animal meeting another of the right species without being eaten must be enormous. Some species overcome this by being hermaphrodite – each individual being both male and female. Sometimes they function as males first, changing into females later; but in others they function simultaneously as both sexes and in exceptional circumstances may even be capable of self-fertilization.

Once fertilized, the eggs of many deep-living species float up towards the surface and the larval development takes place in the highly productive surface layers. The mature larvae migrate down again and metamorphose into the adult form. This vertical migration during their life histories is paralleled by the diurnal migration of many of the species that live by day above 700–800 m (2425–2600 feet).

PAGE 24

This *Argonauta* is an octopus which is often caught quite close to the surface when there are abundant swarms of jellyfish or salps. The argonaut's shell, secreted only by the female, is used as a carriage for her eggs. Her upper pair of arms are specially modified to carry it. The 5 cm long shell is not calcified and so is very light rather like papier-maché. It is quite a different structure from the true molluscan shell. Male argonauts are only 2 cm long.

PAGE 25

Cranchia scabra is [illegible] ely
resembles a goose [illegible]
colour by expand [illegible] hite
pigments cells wh [illegible] er
its body. Its bloo [illegible] of
its body volume [illegible] en-
trations of amm [illegible] s the
animal neutrally [illegible] nimal
like this is a far [illegible] han
its torpedo-shap [illegible] es of
this squid are co [illegible] the
stomach conten [illegible] eir
gelationous ten [illegible] res in
diameter, are a [illegible]
retractable hoo [illegible]

LEFT

Calliteuthis is a squid which lives at depths of about 500 m (1625 feet). This view of its underside shows that, like many fish and prawns living at similar depths by day, the belly is covered with ventral photophores. The left eye of this squid is enormous, while the right one is normal size. The large eye has a yellowish lens which is typical of deep-living animals; while the smaller eye has a clear lens, typical of shallow-living animals. The inference that one eye is used when the squid is deep down and the other when it is shallow is unlikely to be correct since these squid are caught only at depth.

RIGHT

Spirula is a deep-sea relative of squid and octopuses which grows to a length of 7 cm. Towards the top of the body can be seen the outline of its chambered spiral shell. Each sealed compartment is filled with gas at atmospheric pressure. When the animals die and rot away, the shells bob up to the surface and become washed up on shores. Until the 1850s biologists familiar with the shells believed the animals lived inside them. Now we know the shells are internal. Similarly, it is believed that the shells of fossils, such as belemnites, were internal and used as buoyancy organs.

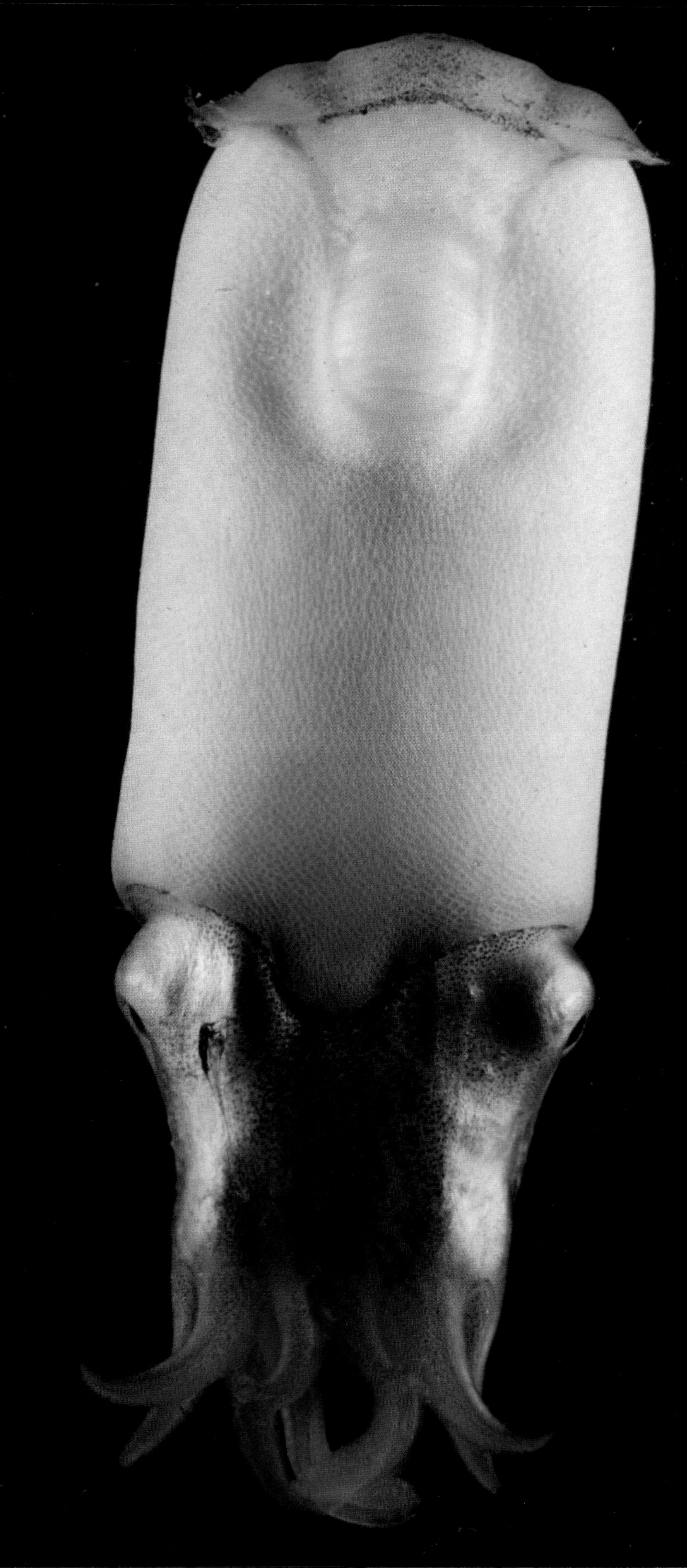

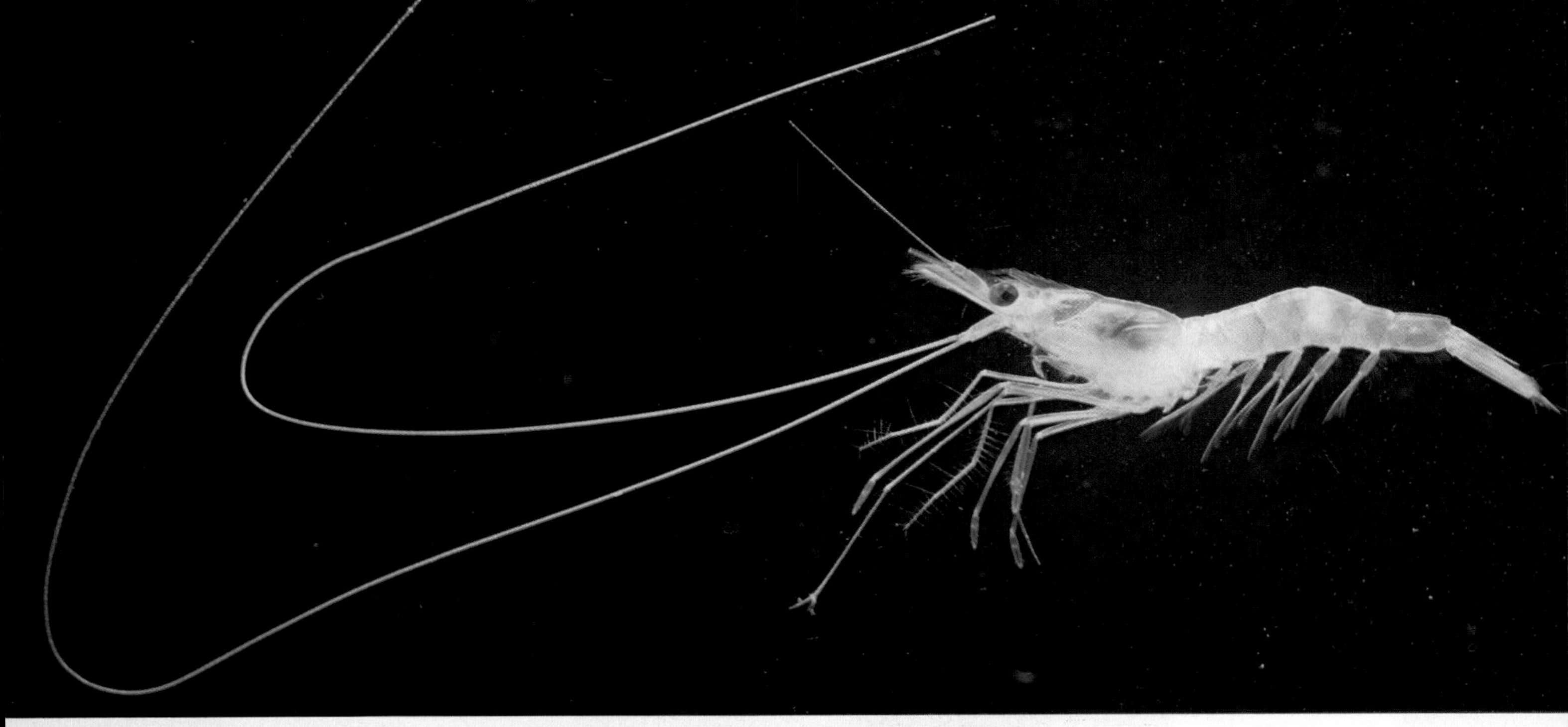

ABOVE
Prawn-like *Funchalia* are found at the same depths that the silvery-sided fishes live by day. They are either half-red and half-transparent, or as in this case, almost colourless. They are active carnivores eating large planktonic animals. In turn, they are eaten by large fish and squid. By day these 10 cm long prawns are caught at about 400 m (1300 feet) but at night they migrate up, close to the surface. Their long antennae are studded with sense organs.

LEFT
A *Pseudoscopelus,* which belongs to a family of fish known colloquially as the great swallowers, attacks a hatchet fish *Argyropelecus*. Its jaws are flexible and extensible and it swallows by using the jaw muscles rather than the gullet muscles as in most other fishes. The drab brown colour is characteristic of fish living below 650 m (2725 feet). This picture emphasises the distinctiveness of the colouration of the two fish each living by day at different depths.

RIGHT
Parapandalus, another type of prawn, lives by day at the same depths as the silvery-sided fishes. Its half-red and half-transparent colouration is characteristic of many of the prawn species from these depths. This 5 cm long female is carrying her eggs stuck to her legs on the underside of her body. She will carry them for several weeks until they hatch into tiny larvae. These larvae live in the surface layers feeding initially on phytoplankton; but then as they grow they switch to a carnivorous diet and eventually migrate down into the depths where the adults live.

LEFT
Benthalbella infans is a mesopelagic fish which has remarkably modified eyes. The lens is spherical and sits on the top of the tubular retina so that the fish looks vertically upwards. Just below the lens is a white structure called the lens pad that enables the fish to see to some extent horizontally. Individuals are hermaphrodite, which means that they are simultaneously functional males and females. The mature fish have four light organs on their bellies, which are exceptional in being formed from modified muscle tissue.

CENTRE
Eutaeniophorus is in many ways a most remarkable fish. It is extremely rare and very few have ever been caught. When caught, the belly is often as full as this bloated animal's. Examination of the stomach contents of one specimen showed that its amazing gut-full was all copepods of one species. Its mouth, much like a sea horse's, has the end turned dorsally. To feed, *Eutaeniophorus* has to strike upwards. In spite of its bizarre character, it is not thought to live deeper than 500 m (1600 feet) or so.

BELOW LEFT
Whale fish are caught below 800 m (2600 feet). They show many features which are typical of deep-living fishes. The eyes are very small while the pupils are relatively large. The fish, which cannot perceive images, is able to see changes in the brightness of the light around it. The mouth is large but the body musculature is poorly developed. This species, *Barbourisia rufa*, is exceptional in being red-coloured. The large pores, opening in a line along the body, mark the position of the extremely large lateral-line canal.

ABOVE RIGHT
The majority of angler fish are black. *Edriolychnus*, however, is sufficiently translucent to be able to distinguish much of its skeleton. It is also exceptional in having its lure modified into an intensely bright-light organ. Since the lure is moveable, this angler must be able to use it very much like a spotlight to flood light its prey. The angler fishes are a very widespread and diverse group below 1000 m (3250 feet). Their larval stages are planktonic, living near the surface. They migrate back down into deep water as they mature. The males of anglers eventually become parasitic on the females.

RIGHT
Caulophryne jordani, like all deep-sea angler fishes, feeds on anything it encounters from copepods and prawns to squid and other fish. Its lure or esca, unlike other angler fish, is non-luminous. Associated with its non-luminous lure is the incredible elaboration of the lateral line system which senses water movements. Each one of the long filaments carries the sense receptors of the lateral line. This horrific looking fish is only 10 cm long. The tiny eyes and large mouth with inwardly pointing teeth are common features of fishes caught below 1000 m (3250 feet).

Whales and giant fishes

The largest animal living today is the blue whale, (*Balaenoptera musculus*), which may also be the biggest animal that has ever lived. The largest specimen recorded was a 36 m (109 foot) long female, weighing well over 150 tons. These giant animals are mammals. Like all mammals they are warm-blooded and bear their young live. At the end of a gestation period of nearly eleven months, a calf about 8 m (25 feet) long weighing nearly 3 tons is born.

In spite of their size, these giant whales do not prey on other large animals but feed by sieving small planktonic animals from the water. In the Antarctic the staple diet is the five cm (two-inch) long euphausiid, *Euphausia superba*, known as krill. Before the whaling industry reduced the stocks of blue, fin and sei whales, it was estimated that each year they consumed 77 million tons of krill. Clearly, these huge mammals can survive only where there is an abundance of food. It seems tragic that these leviathans who so efficiently utilize krill should have been so excessively over-exploited. Man must learn to optimize his use of resources. The whales will be able to recover their former abundance only if given an extensive period of complete protection.

The largest carnivore in the world is the sperm whale (*Physeter catodon*). An average male reaches 15 m (47 feet) long, and weighs 33 tons, but length records exceed 22 m

(65 feet). The females average only 11 m (33 feet) in length. Sperm whales hunt squid – eating even the giant squid *Architeuthis*. They can dive to immense depths. One whale became entangled in a submarine cable off Peru at a depth of 1240 m (3720 feet). Another, followed by a spotter aircraft, dived for nearly 2 hours in a depth of 3493 m (10,476 feet). When it re-surfaced, it was killed and found to contain two freshly-eaten, bottom-living sharks in its stomach.

Many of the smaller whales, such as dolphins and killer whales, are carnivores, hunting their prey by using a sort of asdic system. Reflections of the high-pitched sounds they produce allow them to locate and track down other animals. Dolphins eat fish and squid. Killer whales hunting in packs will attack large whales. They are extremely intelligent hunters, often catching seals in the Antarctic by tipping them off ice floes. This hunting technique makes them particularly dangerous to man. Recently, several yachts have been damaged by killer whales in the central Pacific.

The largest sharks are also plankton feeders, such as the whale shark (*Rhineodon typus*) which grows to over 17 m (50 feet) long and the basking shark (*Cetorhinus maximus*) which grows to over 13 m (40 feet). The whale shark is noted for having an incredibly thick skin; a 10 m (30 feet) specimen has skin 10 cm (4 inches) thick. The largest specimen on record, estimated to be 25 m (75 feet) long, was one frequently encountered off the Yucatan Peninsula.

The manta ray, or devil fish, is another filter feeder. Like sharks and rays in general, it has a skeleton made of cartilage instead of bone like the true bony fishes. The largest species of manta measures 7 m (22 feet) across its triangular pectoral fins and can weigh more than 3500 pounds. The mouth is at the front of the head not underneath as most sharks and rays. From each side of the mouth, a horn curves forward. These horns help to direct the plankton towards the mouth. Manta rays live in tropical and subtropical waters, and because they feed near the surface, they are often seen by skin divers. These huge rays swim through the water by flapping their wings, rather like a bird or a flying fox moves through air.

The carnivorous sharks are the most feared predators of the ocean. The record for a great white shark (*Charcharodon carcharias*) is 12 m (37 feet), which was the length of one trapped near New Brunswick. Fossil teeth that have been found suggest that 27 m (80 feet) long specimens existed in the Miocene. Greenland sharks (*Somniosus microcephalus*) and tiger sharks (*Galeocerdo cuvierii*) can also exceed 7 m (20 feet) in length and weigh nearly a ton. Some of these sharks can swim at great speeds. A mako (*Isurus oxyrinchus*) kept ahead of a launch travelling at 29 mph for half an hour. The fastest

recorded swimmers in the sea are thought to be sailfish (*Istiophorus platypterus*). A speed of 68 mph was estimated for a specimen hooked on a line. Yet the fastest fish over long distances is probably the marlin (*Tetrapturus* sp.) – capable of reaching 50 mph. A swordfish (*Xiphias gladius*) bill, which penetrated 22 inches into a ship's timber, is proof that the fish must have been travelling over 50 mph at the time of impact. These high speeds are made possible by the evolution of a special type of muscle. Rich in the blood pigment haemoglobin, the muscle is red in colour, and the blood-flow within it is arranged in a counterflow system to conserve heat. The muscle temperature of a swimming tuna has been found to be 10°C above the surrounding sea temperature. This raised temperature improves the efficiency of the muscle. As a result, it develops three times the power it would if the fish was totally cold-blooded. Thus many of these sharks and large fish are almost warm-blooded.

Tuna is an important source of protein in many countries. In recent years, there was a big scare when some canned tuna was analyzed and found to contain an appreciable – but by no means dangerous – quantity of mercury. This soon led to the panic withdrawal of tinned tuna for sale. Follow-up analyzes suggest that many of these large predators may naturally have quite high levels of mercury. However, these levels would only be toxic if one ate tuna for every meal on every day for many many years.

One of the bulkiest bony fishes is the ocean sunfish (*Mola mola*). This is a sluggish beast which can grow to an enormous size. In 1908 a steamer off New South Wales rammed one which was 3 m (10 feet) long, 5 m (14 feet) high and weighed $2\frac{1}{4}$ tons. The sun fish is a champion egg-layer with an average clutch of about 300 million.

Inspired by the myth of the kraken, the giant marine animals which always catch people's imagination are the giant squid, *Architeuthis*. The largest specimen on record was 18 m (56 feet) in overall length and, like many others, was washed up on the Newfoundland coast. However, off the Maldives a squid was sighted stretching the length of a 58 m (175 feet) long ship. Intact giant squids are rarely caught. Their anatomy has been worked out from the few specimens washed ashore and from the stomach contents of sperm whales.

The mythology of the sea is rich in tales of monsters and sea-serpents. The size of the oceans and their depth always make it possible that there are large animals still unknown to man. It is important to remember that large animals need large amounts of food to survive, to grow and to breed. If monsters do exist they will be found in the highly productive areas and not in the wide, barren wastes.

PAGE 32
Killer whales, small whales related to dolphins, are extremely widespread and occur throughout the world's oceans. However, it is in the Antarctic that these whales have become notorious. They hunt in packs and are among the most intelligent of all carnivores in the sea. They specialize in following whale factory ships, gorging themselves on the carcases tied alongside waiting to be processed. In this picture the U-shaped blow-hole through which they breathe is clearly seen on one of the whales.

PAGE 33
The arrival of a school of bottle-nosed dolphins to play round the bow of a ship is one of the excitements of a long sea voyage. Swimming by beating their tails vertically, instead of from side to side like fish, these marine mammals can swim at amazing speeds. They have to surface every half minute to breathe and so are restricted to the surface layers of the ocean. They feed on fish and squid, using high-pitched sounds to echo-locate their prey. Usually they are most abundant in highly-productive inshore areas, but they appear frequently in the middle of the most barren oceanic waste.

LEFT
In the Lemaire Channel close to Cape Horn a sei whale surfaces. It is the only rorqual whale that is abundant enough to be exploited. There are about 80,000 living in the Antarctic and about 46,000 in the North Pacific. These populations could be cropped by 5,000 and 3,000 whales a year respectively without reducing their number further. Fin whales are uncommon but their numbers are recovering. Blue and humpback whales are now fully protected but it may be fifty years before they recover anything resembling their former numbers. All these whales feed by filtering large planktonic animals out of the water.

ABOVE
Sperm whales are now the only reasonably abundant large whales; even so their heavy exploitation is making large schools, such as this one, a progressively rarer sight. They spend the summers at high latitudes and migrate into the tropics during the winters, usually calving in the tropics. The bull sperm whale is considerably larger than the cow and grows to over 16 m in length. Sperm whales have remarkable diving capabilities, staying under for up to two hours and possibly diving to depths of 2000 m (6500 feet). They feed almost exclusively on squid but feed on fish off the coast of Iceland. Their stomach contents are our only source of specimens of many large squid species.

BELOW
The giant among sharks is the whale shark, *Rhineodon typus*, which can grow to a length of over 16 m. They are mostly encountered in the tropics and subtropics. In spite of their huge size, they are sluggish and inoffensive. They are filter feeders, using their enormous mouths to funnel in planktonic animals and strain them out onto their gill rakers. An enormous mermaid's purse (egg case) measuring 67 x 40 cm, found off Texas, proved to contain a little whale shark nearly ready to hatch. This 11 m animal, photographed off Australia, has about eighteen remoras attached to its lower jaw.

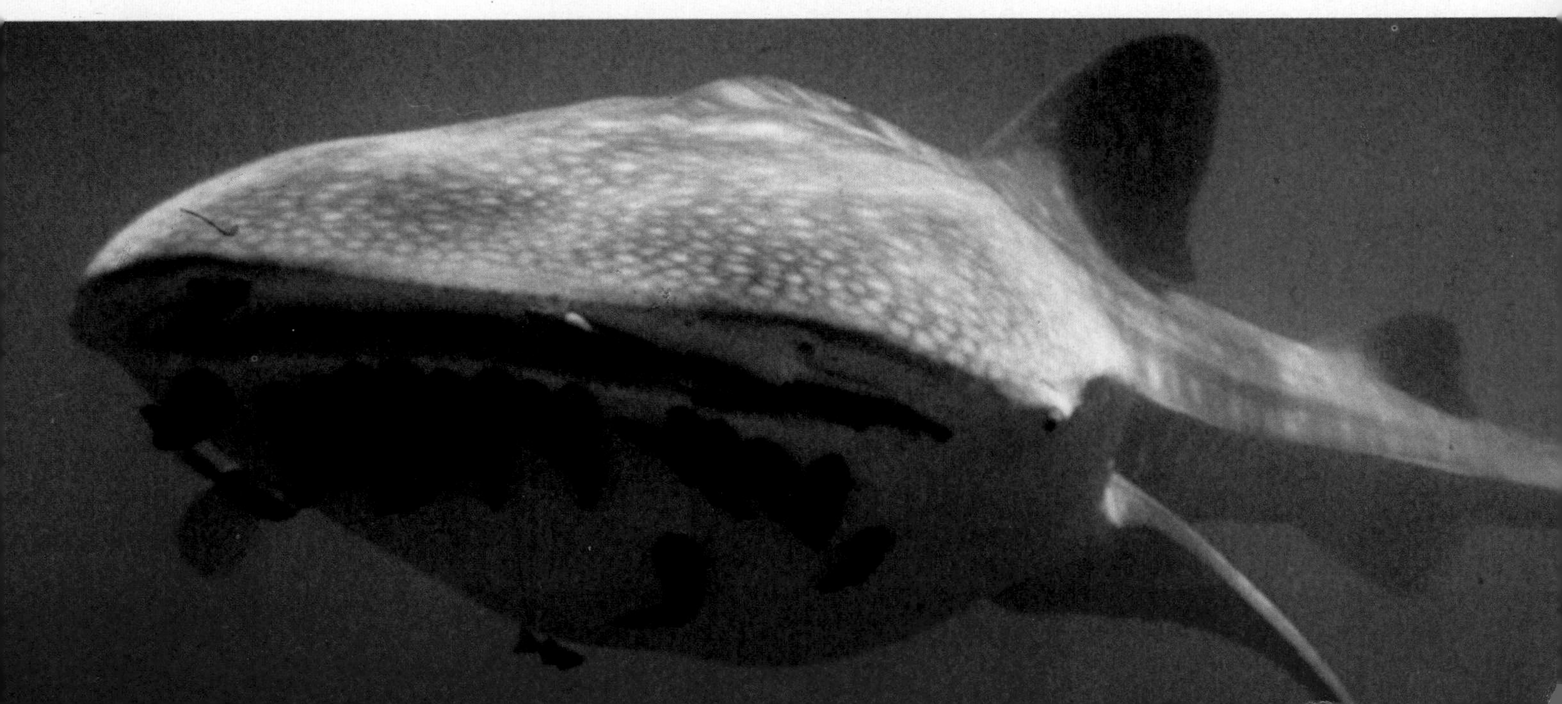

LEFT
Sharks are universally feared as ferocious unfeeling attackers. This picture shows why. The gray nurse shark, *Carcharias arenarius*, is one of the most dangerous sharks in Australian waters and is known to have been responsible for many human deaths. Here, a 2.5m shark shows how devastating an attack by one of these voracious carnivores can be. The huge teeth are ripping into a fifteen pound blue grouper.

BELOW
Photographed through the surface on a flat calm day in the western Indian Ocean, this white-tipped shark, *Carcharhinus longimanus*, is accompanied by a pilot fish and a remora, which is attached just below the shark's dorsal fin. This oceanic shark is frequently seen at the surface following ships in the tropics and is reputed to be a man-eater. Its normal diet consists of fish and squid. They are relatively small sharks. This particular specimen was about 2 m long and when fully grown might reach 3.5 m.

RIGHT
Moray eels have an evil reputation. This is probably because of their snake-like appearance and constantly-gaping mouths.
They are found in the tropics and subtropics and are particularly abundant on coral reefs. By day they hide but at night emerge to hunt out crabs and octopuses. If cornered, they will bite savagely trying to escape. The Romans used to keep morays in fish ponds near the sea and served them as banquet specialities. There is probably no truth in the story that they fed the morays on the bodies of slaves.

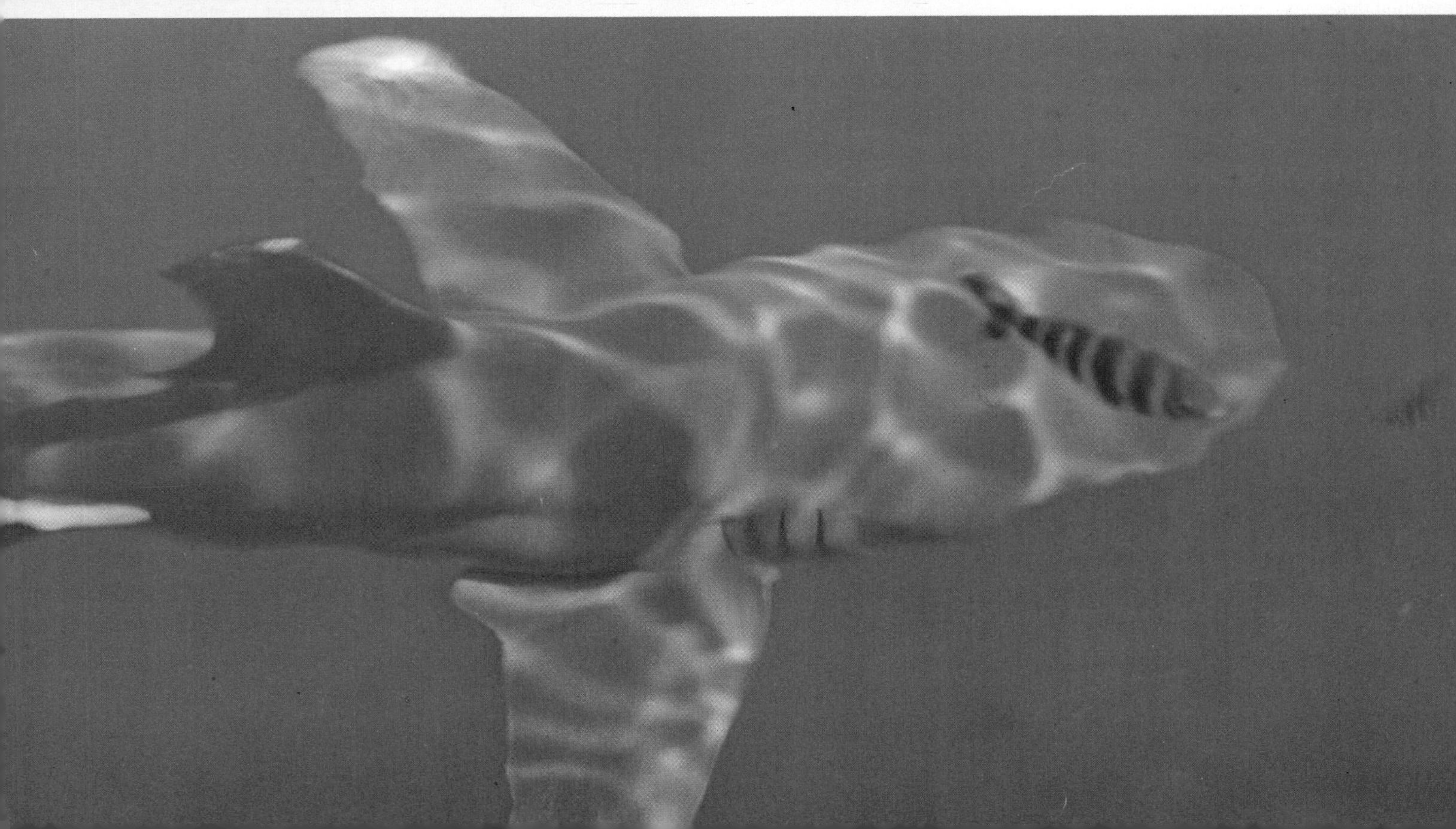

ABOVE
A manta ray glides past a coral head. Its huge, lateral fins beat like wings to drive the ray through the water. Like many large fish, it is a plankton feeder. The big, lateral lappets on either side of the mouth guide the plankton into the mouth. Large specimens are 7 m across and weigh over a ton. They bear their young live. A female, fifteen feet across, bore an embryo 1.5 m across which weighed 20 pounds. The tail is long and slender and in some species carries a poisonous spine.

LEFT
Seen from underneath, this manta shows some of its gill slits which show its relationship to the cartilaginous skeletoned sharks. In contrast to most of the sharks and rays, the mouth is on the front of the head and not underneath. The prominent eyes are keenly perceptive. Despite their alternative name of devil fish they are aimiable, curious creatures. However, they are immensely powerful. When harpooned, individuals have smashed small boats or towed large vessels long distances. Occasionally, they leap out of the water, bellyflopping back with a tremendous crash.

RIGHT
Although it does not grow to a very large size, the electric ray, *Torpedo marmorata*, is still a fascinating animal. On either side of its head are two kidney-shaped organs which are made up of tubular structures each containing 140–1000 electroplates. The electro plates are modified nerve muscle junctions. Small electric rays produce about 40 volts, and the largest six-foot specimens have been measured to give shocks of 200 volts with a current of 8 amps. Since their stomachs have been found to contain large active fish, it is clear that they use their electrical discharges to stun their prey. Other skates and rays also produce very weak electrical discharges but it is not known for what purpose.

Inhabitants of the sea bed

Wherever there is adequate oxygen dissolved in the sea water, there is life on the sea bed. The sea floor is a boundary between the water and outer surface of the Earth's crust. The type of bottom is partly dependent on the water above – its depth, the strength of the bottom currents, the amount of plant growth at the surface and thus the rate at which material is deposited. The sea floor is also dependent on the underlying geological structures and their history. Down the centre of each ocean runs a mid-oceanic ridge, marked by a line of underwater mountains and volcanoes. At the ridge new oceanic floor, which is constantly being formed, pushes out sideways, puckering and creasing the older sea floor so that it either buckles down in the great trenches and is destroyed beneath the continental margins, or pushes the continents very slowly sideways.

Overlying the bottom and not swept clean by strong currents is a layer of sediments. Close to the continents, the sediments will consist largely of sand and dust washed or blown from off the land. They may accumulate at rates well over 1 cm/100 years. Further out in mid-ocean, the only material originating from the land is very fine wind-blown dust; and most of the sediments are the skeletons of surface-living planktonic organisms. Where the surface layers are highly productive, either the silica-like skeletons of diatoms or the calcareous skeletons of pteropods accumulate. Globigerina ooze, the massed calcareous skeletons of *Foraminifera*,

occurs in poorer regions. Once the depth exceeds about 4000 m (13000 feet), the chemical nature of the sea water changes and any calcareous material dissolves. In these areas the bottom deposits are red clays. Here manganese nodules form by the precipitation of metals like manganese, cobalt and nickel round pieces of pumice or sharks' teeth. Red clays accumulate at less than a millimetre per thousand years.

The animals on the sea bed are either filter feeders and deposit feeders or carnivores and scavengers. Filter feeders strain minute particles from the water and so tend to dominate the populations where sedimentation rates are very high. They also dominate in red clay areas where sedimentation rates are exceedingly low. The total weight of living animals per square metre is about 0.1 grammes – about the weight of a flea. Wherever currents sweep away the sediments on the flanks of sea-mounts, great numbers of sea fans, sea lilies, sea squirts, gorgonians and sponges occur. Anchored to the rock, they are orientated so that they are exposed to the maximum flow of water. In this way they are able to filter more efficiently.

The deposit feeders become more abundant where the currents are quieter and the muds accumulate. These sea bed animals, like earth-worms on land, either pick up food particles off the surface or plough through the sediment, swallowing the mud and digesting any food it contains. Some of these animals, such as sea cucumbers, worms and bivalves, can be very abundant and are often extremely patchy. Successive pictures of the bottom, taken only a few metres apart, may show dozens of sea cucumbers in one shot and almost none in the next. This makes it very difficult to estimate the numbers and weights of these animals. One of the most curious of all animals living in the bottom muds is the beard worm or *Pogonophora*. Living in fine, hair-like tubes buried in the mud, beard worms have neither mouth nor gut. Since they only occur where there are very rich, organic muds and the bottom temperature is less than 12°C, the beard worms live by absorbing simple, solute organic compounds in the interstitial water in the mud.

The carnivores are surprisingly abundant, ranging from small crustaceans to quite sizeable fish and sharks. On land and near the surface, the weight of herbivorous animals always greatly exceeds the weight of carnivores. This may not be true, though, at the bottom of the deep ocean. There are thought to be two main sources of food. Firstly, organic substances are get carried off the continental shelves in turbid currents which are great boluses of water made heavy by the mud that cascades down the slope. Secondly, organic material settles in a slow, fine rain from the surface. However, any organic material that is easily broken down will have been removed by bacterial decomposition in mid-water. If these fine sediments are to be of any use, there will have to be some highly effective bacteria at the sea floor. A few years ago Alvin, the American submersible used by the Woods Hole Oceano-

graphic Institution of Massachusetts, was swamped at the surface and sank. Ten months later when it was recovered, the crew's sandwiches, which had been left inside as they abandoned ship, were found in perfect condition. Therefore, the activity of bacteria on the bottom in deep water cannot be very rapid and may not be sufficient to explain the abundance of the deep sea carnivores.

Until recently the deep fauna was studied mostly by trawl or grab samples that were extremely time consuming, technically difficult and expensive. Submersibles, which are very expensive, now enable the scientists to actually see these bottom communities. Two new techniques also make possible a more extensive study of bottom carnivores. The simplest is droplining. Fifty baited hooks or baited traps are set on the bottom and buoyed off for a few hours. The second system is the pop-up camera developed at the Scripps Institution in California. A baited, free-fall camera, which is dropped to the bottom, takes a programmed sequence of still photographs or cine sequences and can then be acoustically commanded to drop its ballast and 'pop up' to the surface for recovery. The speed at which bottom fishes and sharks accumulate around the bait suggests that their main source of food may be the carcases of large whales or tuna fish which sink rapidly to the bottom upon death.

Food must be more abundant on the bottom because, in contrast to the mid-water animals, which are built for economy below 1000 metres (3250 feet), on the bottom it is only below 2000 metres (6500 feet) that similar trends are observed. However, below 1000 metres (3250 feet) bottom fishes also have small and often poorly functioning eyes. They can, therefore, be cine-filmed using bright lights without apparently disturbing them. These bottom fishes have highly developed senses of taste and touch. Their heads are often covered with taste buds. Many species like the rat-tails have long filamentous tails which increases the length of the array of their lateral line receptors. These receptors are sensitive to movements in the water, and the increase in the length of the lateral line organ enables them to locate the source of the movements more accurately.

Many of the bony fishes produce sounds either by using drumming muscles on their swim bladders or by grating specialized parts of their skeletons together. Their swim bladders are also specialized to pick up sounds. In many species only the males have the drumming muscles. Thus the gas-filled swim bladders, instead of primarily functioning as buoyancy organs, may be more important for sound production and reception. The sharks and rays, in contrast, have cartilaginous skeletons and no swim bladders. Even so, they are still very abundant and can grow to quite a size. The primitive six-gilled shark *Hexanchus* can achieve a length of 5 m (15 feet).

To date, a brotulid caught at a depth of 7130 m (23,392 feet) holds the record for the deepest fish capture. Since the Scripps pop-up camera has photographed a fish below 8000 m (26,240 feet), it is only a matter of time before one is recorded in the world's deepest trench 11,515 m (37,780 feet) down in the Mindanao Trench off the Philippines.

PAGE 40
In the deepest part of Sognefjord, Norway, at a depth of 1200 m (3900 feet), a little squat lobster (*Munidia* sp.) emerges from its burrow. At this depth there is no longer any light, so that the light of the flash is the first that this animal will have seen since it was a planktonic larva. In deep water close to land, both the bottom mud and much of the food is washed in off the land. This squat lobster probably feeds by scavenging anything it can find.

PAGE 41
Sea urchins abound in the shallow fringes of the oceans where the shores are rocky but the sea temperature is too cool for coral to grow. Here in the South Atlantic are a group of urchins crawling slowly over the bottom using their suckered tube-feet. They feed by scraping minute algae off the rocks. Experiments carried out by divers to exclude urchins from limited areas show that once the urchins are removed, a lush growth of seaweeds soon appears. The seaweeds soon become grazed down again once the urchins are allowed to return.

LEFT
Each autumn off the coast of Florida and the Bahamas, a remarkable migration takes place. The spiny lobsters form long queues in single file; each animal grasping the tail of the animal in front. The procession may contain as many as fifty animals, both males and females, and it moves off from the shallow water into the depths where they spawn. They seek the relative safety of deep water because mating can only take place immediately after a moult and before the new shell hardens. Initially, the eggs are carried by the females under their abdomens but later they hatch into planktonic larvae.

ABOVE RIGHT
Common inhabitants of the sea bed are various species of spider crabs. This particular species (*Stenorhynchus seticornis*), with its immensely long rostrum, is commonly known as the arrow crab. It is reported that it stores food by impaling it on the rostrum. The arrow crab is widespread throughout the Atlantic in the tropics and subtropics. An almost identical species occurs on the Pacific coast of the Americas. The two species must have diverged on the closure of the Panamanian Isthmus.

RIGHT
Scientists on the east coast of the United States are carrying out an intensive study of the bottom fauna on the continental slope roughly 100 miles south of Nantucket and 200 miles east of New York.
Here, at a depth of 1800 m (5850 feet) a great variety of animals can be seen. There is a large sea urchin (*Hygrosoma petersi*) and several smaller ones (*Echinus affinis*). On the right of the picture are the tentacles of buried cerianthid anemones. Everywhere are specimens of a brittle star (*Ophiomusium lymani*). This picture was taken from the deep research vessel, Alvin, after it had been recovered and refurbished following a ten month period on the sea bed.

RIGHT
On July 19th, 1972 the research vessel, Alcoa Seaprobe, dived to a depth of between 500 and 600 m (1600–1925 feet) in the study area. The bottom fauna was rich and varied. Here a large crab *Geryon quinquedens* adopts an aggressive posture at the approach of the submersible. Beside its legs is a horny tube which is the portable home of a polychaete worm *Hyalinoecia artifex*. The tubes are usually about 15 cm long and provide a useful scale for the crab. The crab is both a scavenger and a carnivore, while the worm is a selective detritus feeder.

ABOVE
On the same dive, the RV Alcoa Seaprobe came across this white hake *Phycis tenuis*. The white hake is a relative of the cod and feeds mainly on small crustaceans and other small fish. Related species are being caught in large numbers in experimental, commercial deep trawling off European coasts.

LEFT
Another exciting moment on the RV Alcoa Seaprobe dive to 500–600 m when two squid are photographed. One squid is defensively producing a cloud of ink; while the other, with its fins and arms outspread, has a dark, transverse band of colour suffusing across its back. Notice how the fine sediment is being stirred up. A *Synaphobranchus* eel swims unconcernedly along in its typical, slightly head-down posture. The eel is near the ceiling of its depth range and lives as deep as 3000 m (9750 feet). It also feeds on small animals. The larval stages of the eel are long, slender and very transparent. These so-called *Leptocephalus* larvae occur abundantly in the surface plankton.

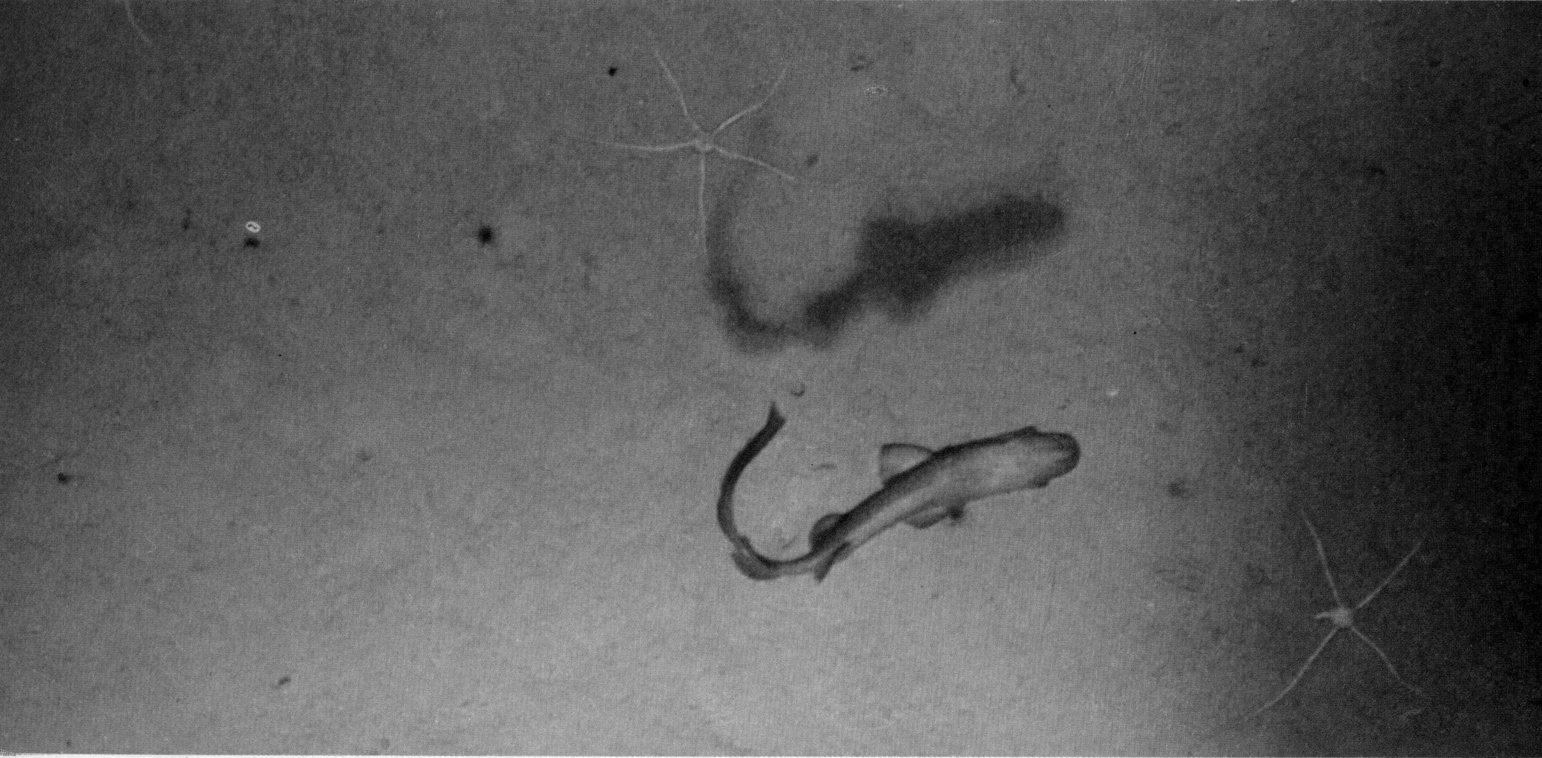

TOP

This portrait study of a rat-tail, or macrourid fish, illustrates the lack of colour of many deep-sea fishes. The tail is long and filamentous. The mouth is under the head and usually these fishes swim with their heads slightly down which is an appropriate posture for a bottom feeder. The large eyes are common to most deep-sea fishes down to depths of about 2000 m (6500 feet); deeper-living fishes have regressed eyes. On the belly is a large light organ. Portuguese fishermen rub rat-tails on their baits, thereby transferring luminescent bacteria to the bait to make it more attractive.

ABOVE

At 1500–1600 m (3900–4225 feet), the RV Alcoa Seaprobe encountered another typical member of the deep bottom fauna – a small black shark. These small sharks are predators and scavengers. The brittle stars, in contrast, are selective detritus feeders, picking up food particles off the mud surface. The Cerianthid anemones, whose tentacles also appear in the picture, feed on small particles that drift down to them. These pictures give an indication of the rich diversity of animals that live on the bottom in the deep sea.

RIGHT

First caught close to the Faroes in the late 1800's in a plankton net haul, *Stephalia coronata* is one of the world's rarest siphonophores. It now seems likely that this plankton net actually trawled along the bottom, since the catch contained several animal species that are bottom-living. *Stephalia* has been caught recently off the north-west African coast and related species have been photographed by pop-up cameras off the coast of California. *In situ* the animal's tentacles trail along the bottom, presumably catching small animals from off the mud surface.

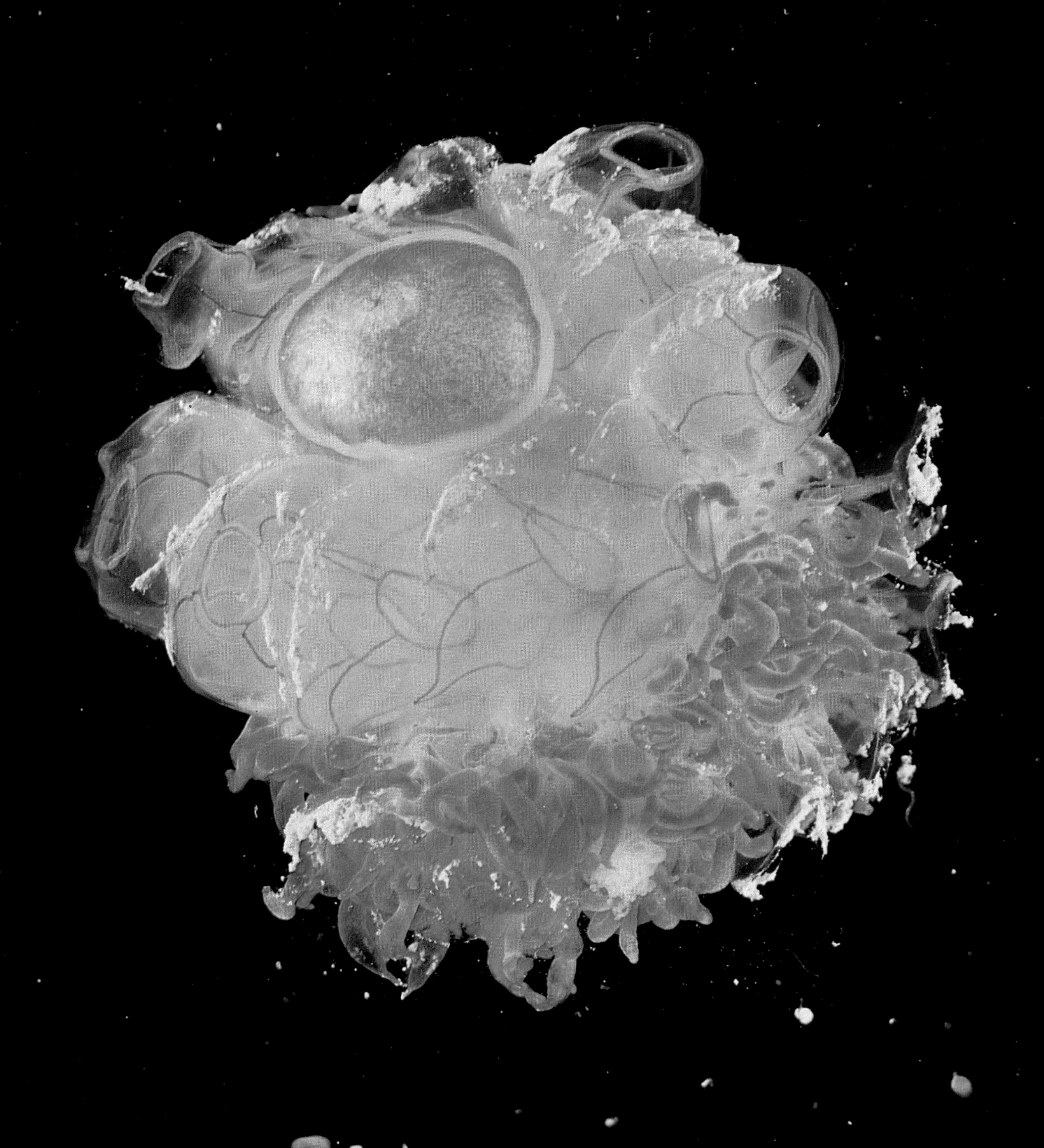

Coral reefs

The most complex habitat in the oceans is the coral reef. The reef-building corals, which are related to sea anemones, secrete their massive calcareous skeletons from the bases of the living polyps. They are restricted in geographical distribution to seas where the temperature never falls lower than 18°C. They are also limited to the surface zone where the plant growth occurs because within the soft tissues of the corals live plant cells called *zooxanthellae*. The *zooxanthellae* are dinoflagellates which are capable of photosynthesis. They get protection and a supply of excretory compounds from the coral, and in turn provide extra oxygen for their hosts. Living together with mutual benefit is called symbiosis. Another example of a symbiosis is the *zooxanthellae* that live in the tissues of the giant clam *Tridacna*. The successful relationship between animal and plant possibly allows certain species of *Tridacna* to grow over a metre in length and to weigh up to 200 kilos.

Mostly nocturnal, the coral polyps emerge from their skeletons at night to catch small particles and zooplankton from the water. Delicate and unable to withstand mud and sand settling on them, corals are very sensitive to any form of of pollution that increases the amount of sediment in the water.

In the Indo-Pacific region, there are over 700 species of reef building corals, but in the Atlantic there are only about 35; this remarkable difference reflects the geological history of the two ocean systems. The richest reefs occur around the equatorial atolls in the Pacific area and along the Great Barrier reef,

which runs over a thousand miles down the northeast coast of Australia from New Guinea to Queensland.

Coral reef formation is restricted to clean water of the surface 30 m (100 feet). Thus it may at first be surprising to learn that coralline rocks are found capping underwater sea mounts at much deeper depths. However, the sea floor may be rising or falling as a result of the forces that cause continental drift. In addition, during the Ice Age the sea level was much lower. Deep bore holes made at Eniwetok Atoll showed that the atoll sits on top of an ancient volcano capped with nearly a mile of coralline rock – the result of tens of millions of years of coral growth.

The rate of coral growth depends very much on its type. In good conditions the stagshorn types grow 2–5 cm (1–2 inches) a year. The massive brain corals grow more slowly. The coral is attacked by a host of animals, from boring sponges worms and bivalve molluscs to crown-of-thorn starfish and fishes such as the parrot fish. Wave action breaks up the coral skeletons so that the pieces accumulate in areas of debris. The debris may either be broken down further to sand by being eaten by sea-cucumbers or else cemented together by encrusting calcareous algae. On reef crests, where wave action is maximum, about half of the calcium carbonate is derived from these plants rather than from the corals.

The type of corals vary in different places on the reef. Typically, a fringing reef consists of an inshore reef platform with the reef crest beyond it. From the crest the reef slopes steeply into deep water and the bottom of the reef is marked by a tallus slope of coral debris. Small hemispherical branching colonies inhabit the reef platform where they have to withstand intense sunlight, low oxygen levels at night and flushing with freshwater during tropical storms at low tide. The reef crests are inhabited by wave resistant forms, such as the stinging coral, *Millepora*, and large areas are covered by calcareous algae. Stagshorn and columnar forms are found just below the crest but near the top of the reef. Massive forms, like brain corals, are deeper down or in the calmer turbid water in sheltered lagoon areas. Deepest of all are gorgonians and delicately branching and bracket forms. This zonation is not always clear as the deep water forms may occur at shallow depths in deep gulleys.

Soft corals are abundant on Indian Ocean reefs such as those off the Seychelles. Although they have hard skeletons, the stinging corals (*Millepora* spp.) are related to the soft corals. These and other related groups, such as the gorgonians, are much more highly-coloured than the reef builders.

This abundance of species and coral forms results in a multitude of microhabitats. Each microhabitat tends to have a special fauna associated with it. As a result, there is not only a zonation of the corals but also of the fishes and other animals. Surgeon fish like *Acanthurus triostegus* are herbivores and occur in shallow water, whereas the little brightly coloured

coral fish (eg *Pomacentrus* and *Abudefduf* species) are omnivores and are more widespread. Parrot fishes have horny beak-like teeth for rasping coral growths.

Brittle stars, worms, sponges and sea anemones live on and between the coral branches. Boring species of sponges and bivalve molluscs penetrate into the coral skeletons. Crabs also live in the coral, causing it to produce gall-like growths. Live sponges, sea-squirts, snapping shrimps, cowries, cone shells, worms and sea-cucumbers live underneath the coral. Often, each species feeds on a very narrow range of food. Some cowries, for example, feed only on a single species of sponge. Some cone shells eat other molluscs, and some even predate on fish, killing them with poisonous darts. They are also dangerous to man.

There are some remarkable associations between different species. Clown and damsel fishes dance in and out of the tentacles of the big sea anemones. Their gaudy colour warns off possible predators, and displays to others of their own species possession of this particular anemone. Cleaner fishes and shrimps are also strikingly-coloured. Trying to attract other fish they perform dances. A host fish, responding to the cleaner's dances, hangs motionless in the water, allowing the cleaner to eat any external parasites, even letting them clean off the gills. Another little blenny mimics the cleaners, but instead of cleaning the hosts bites lumps out of them.

Care is needed when exploring reefs as a surprising number of the animals are armed with poisonous spines or stings. Most common are the armies of black sea urchins (*Diadema* sp.) with long slender spines which are tipped with poison and break off in the wound. Stinging coral and some sea whips sting painfully. More dangerous are the sting rays which hunt for food on the sand around coral heads. The gaudy lion fish, which has toxic spines, exhibits its danger, whereas the stone fish lies superbly camouflaged on the bottom. The stone fish's sting can be fatal. Sometimes the danger is totally unexpected. Ten years ago, the blue-ringed octopus, common on some Australian reefs, was thought to be harmless. Now this animal, which has a spread of 10–15 cm (4–6 inches), is known to have a bite that has killed several people.

At night the reefs become exciting, dangerous places. All the predators emerge from their hide-outs and scour the reef for food. Moray eels hunt for crabs, sharks move up from deeper water and crayfish prowl along the bottom.

Types of coral do occur in deep water. The precious coral, *Corallium rubrum*, found in deep water in the Mediterranean, is a gorgonian. Cold water reefs of *Lophophyllia*, associated with the furrow marks gouged by icebergs during the Ice Age, occur in the Norwegian Sea and off the west coast of Scotland.

PAGE 48
Off the coast of Mozambique, a shoal of transparent juvenile fish shelter among the the branches of a stagshorn coral (*Acropora* sp.). Around them swim more intrepid blue coral fish (*Pomacentrus* sp.). Stagshorn coral grows in clear water where the wave action is not too heavy. The branches grow extremely rapidly, as much as 10 centimetres a year. The maximum growth rate measured in a coral is 25 centimetres in a year by a coral living off Barbados.

PAGE 49
This is part of a colony of a soft coral *Dendronephthya* sp. It is an alcyonacean octocoral, typically having eight pinnate tentacles on each polyp. Unlike the reef-building corals whose skeletons are external, the octocorals have internal skeletons. The calcium carbonate spicules inside the soft tissue show up clearly. These alcyonaceans dominate the reef-dwelling, soft corals in the Indo-Pacific but are absent from the Atlantic reefs. In many species the polyps actively comb the water, often with all the polyps of a colony synchronized together.

LEFT
Gorgonians are the dominant reef-dwelling octocorals in the Atlantic. However, they are also common in the Pacific, especially towards the deeper parts of the reefs. They do not contain the symbiotic algae that occur in the reef-forming corals and so flourish at considerable depths. Also, the soft corals tend to be much more brightly-coloured. The skeletons can be either calcareous or horny. Many species are much more simple than this species with its intricate tracery of branches.

TOP RIGHT
The slanting rays of sunlight illuminate the tips of a columnar coral *Acropora palifera* growing off Praslin Island in the Seychelles. The successful growth of coral is dependent on sufficient sunlight, since within the soft tissues of the coral are plant cells called *zooxanthellae*. The plant cells get protection from the coral and in turn provide the coral with extra oxygen as a result of photosynthesis.

RIGHT
Corals are relatives of jellyfish and sea anemones. Here, a species which is unusual in opening and feeding by day shows its similarity to sea anemones. At the centre of each individual polyp is a simple mouth surrounded by rings of tentacles. Most corals are complex colonies of these polyps and they feed by trapping minute, planktonic animals. Their skeletons of calcium carbonate are laid down at the bases of the polyps. Massive forms like this species grow relatively slowly, whereas the more delicate branching types grow more rapidly.

ABOVE
110 miles above the South Pacific the astronauts in Apollo 7 looked down on the Tuamoto Archipelago. In 1835 Charles Darwin sailed among these atolls in the Beagle and began to evolve his theory of the formation of atolls on subsiding volcanoes. Recently, borings at Mururoa in the Tuamotos have shown that the atolls are subsiding at 2–3 cm/1000 years. However, the coralline limestone has been accumulating at 2 metres/100 years, keeping up with the rise in sea level of about 100 m in the last 15000 years.

ABOVE LEFT
A great shoal of a surgeon fish (*Acanthurus triostegus*) swim past the camera in the Indian Ocean. At the base of the tail of each fish is a large sharp spine, which gives them their vernacular name. These fishes are always found in shallow water, since they are vegetarians feeding on algae which seldom grow deeper than 5 m (15 feet). Many reef fishes with this pattern of vertical banding are sociable and congregate in sizeable shoals. This species is particularly abundant around the Hawaian reefs.

CENTRE
Rock cod are common members of the reef fauna, but normally by day they skulk away in deep recesses, emerging at night to feed. The blue-spotted rock cod *Cephalopholis miniatus* is a particularly handsome species and is widespread throughout the tropical Indo-Pacific. Although this species only grows to a length of half a metre, others grow to four or five times this length. Unfortunately, spearfishermen have hunted out and indiscriminately killed many of these handsome monsters.

LEFT
A school of two species of Gaterins (*Gaterin reticularis* and *G. gaterinus* with the yellow tails) glide past a coral head off the Kenyan coast near Malindi. Shoals like these were commonplace when skin-diving first began. The invention of the speargun meant that edible fish like these have become very nervous. Since the complex fragility of the coral-reef fauna has been recognized, many enlightened governments are setting up marine parks where the animals are completely protected.

BELOW
This delicate and beautiful little fish (*Anthias anthias*) was photographed among a scarlet forest of gorgonians on a Red Sea reef. Coral reefs, because of their variety and complexity in shape and form, present a complex environment containing many microhabitats. The richness of the habitat has resulted in remarkable evolutionary radiation of the associated fauna. The reefs of the Indo-Pacific are much richer in their variety of species than the Atlantic reefs. The reef fishes in particular are more spectacularly coloured than any other group.

BELOW RIGHT
The zebra angel fish (*Pomacanthops* sp.) swims moodily past a coralline rock face of an Australian reef. They are solitary fish which have territories and drive off other fish of the same species. They feed by browsing on minute animals and plants that cover the rocks. In the background hover a group of soldier fish (*Holocentrus sp.*) These are much more sociable fish but they are also much more wary, usually retreating deep under overhangs at the approach of a diver.

BOTTOM
A rainbow parrot fish (*Scarus guacamaia*) rests within a recess in the coral. Parrot fishes have strong, horny teeth with which they rasp at the coral as they feed. They are excellent eating and so are a favourite prey of spearfishermen. In front of the parrot fish is a sea urchin with long slender spines. These urchins (*Diadema* sp.) also have scraping teeth on the undersides of their round bodies. Although they have no proper eyes, they can sense shadows and furiously wave their poison-tipped spines at the approach of any danger.

LEFT
With its two large fluted shells buried within a crevice in the coral, this giant clam *Tridacna gigas* reveals the elaborate folds of the edge of its mantle. Within these blue and green folds symbiotic algae, or zooxanthellae, are concentrated. The clam is a filter feeder, straining fine particles out of the water; however, the zooxanthellae seem to enable the clam to grow to an enormous size. Along the edge of the mantle the row of black dots are simple eyes; any shadow passing over the clam causes it to clamp shut.

ABOVE
Feather stars, or crinoids, are some of the most delicately-coloured and fragile of all the reef animals. The central disc is attached to the substrate by claws; in some deep-sea forms it is mounted on a stalk. From the disc radiate five arms which soon subdivide into numerous pinnate branches. The arms curl and uncurl in the water currents, trapping minute food fragments which are carried wrapped in mucus along a longitudinal groove on the arm down to the mouth in the centre of the top of the disc. These strange relatives of starfish have a long fossil record stretching back 400 million years to the Ordovician Age.

LEFT
Coral reefs are rich in all types of echinoderms, which include sea urchins, starfish, brittle stars, sea cucumbers and feather stars. Sea urchins are notorious for their long, poisonous spines. Not all of them are quite so unpleasant. This slate-pencil urchin has thick, heavy, triangular spines. This ten centimetre specimen was photographed on a reef flat off Mozambique. These urchins are grazers, rasping algae off the surfaces of the coralline rocks. The spines of slate-pencils urchins are often used as decorations and in the manufacture of aeolian instruments. When the wind blows on them the spines bang together, making a tinkling noise.

ABOVE RIGHT
The crown-of-thorns starfish (*Acanthaster planci*) is a great scourge of coral reefs. A few years ago, it was considered to be uncommon, but with the spread of skin-diving, it was found to be abundant. Its

abundance was attributed to a population explosion caused by man's interference with the underwater environment. Swarms of the starfish were found crawling over reefs, eating the soft tissue of all the coral they encountered. It was feared that they threatened the very existence of the reefs. However, this is no longer considered a serious danger. The starfish with its poison-tipped spines is not an animal to tamper with.

RIGHT
This attractive, blue-ringed octopus (*Hapalocheana maculosa*) is a common inhabitant of Australian reefs. Its body is only two to three centimetres long. Yet, it is one of the most dangerous animals in these waters. It is only recently that the bite of this tiny animal was found to be lethal. It can cause more deaths in a year than notorious sharks. Normally, this octopus feeds and attacks moving animals, but if handled will bite with its parrot-like beak, which is situated in the centre of its eight arms.

BELOW RIGHT
A live tiger cowrie (*Cypraea tigris*) shows its spots as it crawls over a brain coral. The edges of the mantle completely cover the shell for most of the time, keeping it shiny and free from encrustations. At any hint of danger, the mantle folds back showing the shell. The shell colour in many cowries is a warning that if attacked, they can release strong solutions of sulphuric acid. Cowries have been decimated by the ardent and destructive efforts of shell collectors. In many countries it is now an offence to collect any shells – even empty ones.

FOLLOWING PAGES
Amphiprion percula is one of the little damsel fish species that lives in association with some of the big sea anemones that occur on coral reefs. They never stray very far away from their host anemone, a pair of fishes usually guarding their anemones from others of the same species. At the approach of danger they perform an undulating dance above the anemone, finally plunging into the tentacles. They appear to be totally immune to the anemone's stinging cells, whereas other fish species are killed and eaten by them.

Camouflage and display colours

Animal colouration, except when man has interfered by by selective breeding, is highly adaptive. Often its function is to make the animal less obvious to its potential attackers by various forms of camouflage. Conversely, some animals want to appear conspicuous either because they must display themselves to their rivals in order to defend their territory or to show off their nastiness.

In the oceans, colouration is nearly always protective. As discussed in Chapter Three, most of the neustonic animals associated with the water surface are blue. The blue colouration is either a protection against the harmful effects of the ultra-violet light in sunlight, or it is a camouflage against possible aerial predation by oceanic sea birds. Many of these surface animals, if kept in a ship's laboratory, tend to lose their blue colour, slowly turning pinkish. The blue colour in the majority of species is produced by a chemical complex of a protein molecule joined to a red carotenoid pigment. The complex is unstable; and if the animal is unhealthy, its blue colour slowly reverts to the reddish colour of the carotenoid. In the goose barnacle although the main body colouration is not blue, the ovaries in the stalk are invested in a sheath of blue tissue.

Transparency is common among the smaller planktonic species near the surface, but some quite sizeable animals use this method of camouflage. The amphipod *Cystosoma* grows

to about 15 cm (6 inches) in length, and apart from any pigment left in its gut from its last meal is so transparent that it is impossible to see in a plankton sample. To be transparent the animal must be the same optical density as sea water – we still do not know how this is achieved. In many species this needs constant effort because the slightest damage will result in the wound area becoming white and opaque. Some organs are impossible to make transparent, such as the eyes and the gut. Often these are masked with a covering of red pigment or by silvering the outside of the organ.

Near the surface the fishes tend to be countershaded with dark backs, bluish flanks which are often banded and pale silvery bellies. From above and from the sides, the dark back and bluish flanks are effective camouflage, as any skin diver will know. From below, the pale belly is effective until the fish becomes silhouetted against the bright circle of light emanating from the surface. This circle of light is produced by the refraction of light at the water surface, and the fishes do not seem to have evolved the answer to this silhouette problem near the surface.

Below 250 m (800 feet) the fishes use their ventral light organs to break up their silhouettes. Their backs are still dark or black but their flanks are silvery and mirror-like. There is a symmetrical fall-off in light intensity with the brightest light coming from the surface and the dimmest being back-scattered from below. A mirror suspended in the water will reflect the same quality and intensity of light as the background against which it is seen. It will therefore be invisible. The silveriness is produced by platelets of guanine crystals. Each platelet is is quarter of a wave length of blue-green light thick and the same distance apart. At night these mirror sides could become a liability as any flash of bioluminescence (production of light by living organisms) will light the fish up like a searchlight. Most species rely on their greater activity to escape, but one one little fish *Valencienellus* has black pigment cells which can expand and draw a 'curtain' over the mirror sides.

Many squid and crustaceans have similar pigment cells which expand and contract. Sometimes these are under hormonal control so their response is relatively slow. In squid and octopuses the pigment cells are expanded by muscles under nervous control, and colour changes can flush rapidly over the body.

Between 250–650 m (800–2725 feet), the prawns tend to be half-red and half-transparent as in *Parapandalus*, below this depth their colour is totally red. In *Notostomus* the red colouration is the result of a carotenoid pigment being concentrated in the animal's outer skeleton. Since the animals are unable to synthesize carotenoids, the source of the pigment

must be in the diet and originate ultimately from phyto-plankton in the surface layers. Since there is no red daylight at these depths and the red pigments absorb blue-green light, it is an effective camouflage.

Fish also show changes in colouration below 650 m (2725 feet) where there is not much daylight left, and the intensity of bioluminescence becomes relatively bright so that silveriness is no longer an effective camouflage. Here the fish are uniformly coloured a matt black or bronze.

At really great depths both in mid-water and on the bottom, animals are colourless because of the almost total absence of light. The decapod crustacean *Polycheles* is a typical example of a bottom animal; and its lack of colour has a counterpart in the fauna of caves where again there is no light.

On the bottom in shallower depths, colour is important and the primary use is for camouflage. Bottom fishes and prawns have a mottled colouration and the ability to change colour to fit in with their background. The plaice, for example, has the habit of flicking a small amount of sand over the body so that its outline is obscured. The stone fish, the dangerous venomous fish of Australian reefs, has ragged fins and small projections over the surface of its body, which help to break up its outline and give it a non-fish like appearance.

Mimicry of the background is shown by many of the animals associated with the *Sargassum* weed. Not only does the colour match the weed, but also many species have little projections over the body to confuse its outline. Even the little shrimps, which live in close association with corals and sea anemones, mimic the colouration of their host and often have transparent parts to their bodies. This gives the shrimp a very unshimp-like appearance, making it exceedingly difficult to pick out.

Sometimes animals want to be clearly seen and have the appropriate colours. The lion fish *Pterois* exhibits itself, especially when it is disturbed. It swims slowly away with its dorsal and pectoral fins held out stiffly – the gaudy banded colour proclaiming the nastiness of its poisonous spines. Box and puffer fishes are poisonous to eat and some others may also taste revolting.

Exhibitory colour is often associated with distinctive behaviour patterns. The cleaner fish reinforces its colour pattern by a striking dance. The blenny that mimics the cleaner not only mimics its colouration but also its dance. The exhibition of colour is not always towards other species, but some reef fishes exhibit themselves to members of their own species; because like birds in a garden they have territories. Their display and aggressive behaviour is directed towards their own species. In the butterfly fishes (*Chaetodon* spp.), the eye is often camouflaged by a dark eye stripe, and a large round eye spot near the tail may distract any attack directed at the delicate eye to the less vulnerable tail.

Shoaling fishes apparently use their colours to keep the shoal together. Reef fishes can be categorized by their colouration into those which shoal, swim in pairs or are totally solitary. Even an apparently simple pattern of colouration, much of which we still do not understand or appreciate, may have considerable significance in the biology of animals.

PAGE 58
Near the surface countershading is extensively used as a method of attempting camouflage. Many fishes have dark backs, bluish flanks and pale bellies. This picture of a shoal of sardines in the Red Sea shows the limitations of this colouration. The fish directly overhead are silhouetted against the circle of bright light produced by the refraction of light at the water surface. Beyond the bright circle, the fish are harder to distinguish, but the dark backs look paler than the bellies because they are more strongly illuminated. Seen from the side or from above, the countershading is very effective and the fishes would be hard to see.
PAGE 59
This amphipod, *Cystosoma*, is one of the most transparent organisms in the sea. A large 15 cm (6 inch) specimen can be seen only in a net haul by the apparent hole it makes among the other animals. The only colour in this specimen is the interference colour of the eyes and some red pigment from the animal's last meal. How animals make themselves transparent is not understood. It requires effort to remain so, as many animals, if they are damaged or sickly, start to turn opaque. *Cystosoma* is usually caught at depths of about 300 m (975 feet).
RIGHT
This sand smelt *Antherina presbyter* shows countershading beautifully. The back is darkly coloured, whereas the fish's belly is pale. The flank has a greenish tinge below the horizontal black band. In natural lighting the dark band is intensely silvery and this illustrates how the apparent colour patterns of fish can alter according to the lighting. This 15 cm (6 inch) long fish forms compact shoals. The shoals are kept together visually; the silvery band flashes as the fish turns, signalling the manoeuvre to the others in the shoal.
LEFT
Any net hauls, taken by day from depths below about 700 m (2425 feet), are characterized by black-coloured fish and superb scarlet prawns. This specimen, *Notostomus longirostris*, is a magnificent 12 cm (5 inch) example of a scarlet prawn. The red pigment is concentrated in the carapace. It is a carotenoid, astaxanthin, which animals are totally unable to synthesize and so must be derived from pigments in the diet ultimately originating from phytoplankton. The red pigment hardly reflects either the blue-green daylight which penetrates to these depths or the blue-green light of most bioluminescence.
RIGHT
In contrast to the rich colouration of the scarlet prawns captured in midwater, this bottom-living prawn, *Polycheles* sp., is nearly colourless. Daylight is totally absent from the depths at which it lives. Thus colour to these animals is of no consequence. *Polycheles* is totally blind. Observations from submersibles near the bottom have shown that below 1000 m (3250 feet) animals do not appear to notice lights. Thus they can be watched and filmed behaving normally. Cave-dwelling species, which live in conditions of perpetual darkness, are likewise colourless and blind.

ABOVE
Here, a little shrimp, commensal on a coral, mimics the colouration of its host. It has used another technique of breaking up its outline so that it does not appear to be a shrimp without careful examination. Parts of the body are almost completely transparent. There are a great many species of similar shrimps that live in close association with a great variety of hosts, from anemones to sea urchins and sea cucumbers. Many of these have developed very similar camouflage systems.

LEFT
A grouper being cleaned. Normally, the grouper hides away by day in deep recesses in the reef. Its colouration of light and dark bands breaks up its outline so that it is difficult to make out in these dimly lit holes. The irridescent blue spots are often associated with solitary living reef fishes. The colour of the cleaner fish is quite different with its longitudinal blue and black bands. A cleaner fish must attract its hosts so that it can feed on their ectoparasites and not get eaten itself. It advertises that it is useful and not to be eaten partly by its colour and partly by performing a special dance routine.

RIGHT
One of the basic principles of camouflage is to appear to be what you are not. Mimicry is one effective method of achieving this and such mimicry is particularly common amongst the *Sargassum* weed community. This little triggerfish is not only coloured identically to the weeds (even the dark markings resemble the weed's shadow patterns), but the covering of skin processes breaks up the outline of the fish. Even the anemone at the top right of the picture blends in almost completely with the weed.

LEFT
This magnificent emperor butterfly fish, *Pomacanthus imperator*, photographed in the Seychelles, is a glorious example of a territorial reef fish. Each fish occupies its own area of reef which it defends from incursions by others of its own species. Its colour is a signal to keep away or else. However, it has still evolved a black eye stripe so that its most vulnerable and probably its most essential sense organ is not immediately obvious to an attacker. Juvenile fish are a uniform black, and only as they mature sexually do they occupy territory and develop this handsome colouration.

RIGHT
The lion, or butterfly fish *Pterois*, is one of the most spectacularly grotesque fishes of coral reefs. This Australian specimen taken by flashlight flaunts its red-banded colouration and plumed fin rays. It is a fish which is clearly not camouflaging itself but instead exhibiting its unplesantness. The ends of the lion fish's fin rays are tipped with poisonous points. Black and yellow banding is one of the more common colour combinations which are used to warn off other species. The success of warning colouration is illustrated by the number of harmless species that mimic these colours and so get away without being attacked.

LEFT
This butterfly fish (*Chaetodon* sp.) shows how many fish hide their eyes. The round shape and the tendency to reflect light from the eyes is clearly shown by the fish in the background. Butterfly fish are territorial and so display to other members of their own species. Thus much of the time the fish needs to appear obvious. The eye is both a valuable sense organ and is often used as a focus of attack. The broad black band in this *Chaetodon* effectively hides the eye outline. Such eye stripes are common in many other marine and also terrestrial animals.

RIGHT
A close-up of the eye of the long-nosed pufferfish, *Canthigaster margaritus*, which lives in the tropical parts of the Indian and Pacific Oceans. In the 15 cm (6 inch) long specimen, the eye is camouflaged partly by the concentric red rings, and partly by a large black eye-spot situated just below the dorsal fin. The animal's generally gaudy colour is a warning against its highly toxic flesh. The puffer fish further tries to deter any attacker by inflating itself and croaking loudly.

Living light

Many people have had the experience of rowing on the sea or walking along the seashore on a dark night and seeing blue-green flashes of light. The ability to produce light is much more widespread among marine organisms than among inhabitants from terrestrial or freshwater environments. It is a cold light produced by the interaction of two chemicals synthesized by the organisms. Some animals, which apparently produce bioluminescence or living light, do not produce the light themselves but culture luminescent bacteria in special glands. In the lures of some deep-sea angler fish, the glands are sealed with no opening to the outside, and so the bacteria apparently must be passed on from one generation to the next within the fish's eggs.

While luminescent phytoplankton is confined to the surface layers, luminescent animals exist at all depths. Organisms

which luminesce either release chemicals into the water as a cloud or else retain the luminescence within special light organs or photophores. Photophores are often very elaborate structures fitted with reflectors, diaphragms and lenses. How the luminescence is produced depends very much on its function. Clouds of luminescence tend to be produced for defence against predators. The mid-water squid *Histioteuthis* can produce a cloud of either dark coloured ink or luminous ink. The big red prawns *Acanthephyra*, which occur by day at a depth of 600 m (1950 feet) and at night migrate up near the surface, produce massive clouds of luminescence. The planktonic ostracod *Cypridina* also luminesces. It releases the two chemicals from separate glands near the mouth. One substance, called luciferin, in the presence of the other, which is an enzyme called luciferase, combines with oxygen to release the light. No reaction takes place when the chemicals are dried. During the Second World War, Japanese officers used the light given off by moistening dried *Cypridina* to read their despatches at night. Some phytoplankton, like the dinoflagellate *Noctiluca*, produce brilliant flashes of light, possibly to illuminate animals grazing on them so that the animals are seen and attacked by predators.

Lantern fishes have patterns of photophores on their flanks which are species specific. In other words, purely on the basis of the pattern of these light organs, the fish can be identified. It therefore seems likely that the fishes themselves recognize members of their own species by their photophore pattern.

Many of the squid, prawns and silvery-side fishes, which live by day between 250–700 m (800–2425 feet), have elaborate photophores in rows along their bellies. The silveriness of the

fish is an effective camouflage from predators attacking from the side but not from directly below. If viewed from this direction, the fish are silhouetted against an area of brightest light. The glowing photophores on the underside therefore, help to break up the fish's outline. The photophores have to glow at the right intensity if the camouflage is to be effective. Some species have a little light organ inside the eye which can be used as a standard to compare with the light coming from above. Thus the light output of the ventral photophores can be adjusted appropriately.

Male lantern fish have large photophores on their tails. At night, net hauls have been found to contain mainly female specimens of some species, whereas tuna stomach contents in the same area contain mostly males. A possible explanation is that, if a shoal is attacked, the females remain quiet in the water, while the males dash off, flashing their tail photophores and sacrificing themselves to lead the attacker away. The net, of course, is blind to the males' sacrifice and catches the females instead.

'Headlamp' photophores are quite common. These are large light organs on the fish's snout or cheeks which shine forward, probably switched on for the final snap as the fish catches its prey. The exception to blue-green luminescence is in the fish *Pachystomias*, which has big cheek photophores producing red light. Its eyes are also exceptional in being sensitive to red light. It must be ideally suited for hunting red prawns, since red animals can be seen only if red light is reflected from them. *Pachystomias* produces the red light and is able to see it. The red prawns, on the other hand, are unable to perceive the red light and must be easily caught by the fish.

Many types of deep-sea fishes have lures with which they entice their prey nearer, often into their mouths. At night or at great depth, the lure will be of little use if it cannot be seen in the dark, so the lures are often luminous. The lure may not only attract prey but in the case of the angler fishes may also attract the males to the females.

Another possible function of light organs is make the animal look bigger than it really is. Some crustaceans have light organs on long spines or processes at the extremities of their bodies. In this way an animal with a body only a centimetre long may have light organs spread over two centimetres apart. Similarly, the viper fish *Chauliodus* has a thick gelatinous sheath covering the whole of its body. Within this sheath are hundreds of minute light organs which are best seen by their red fluorescence in ultra-violet light. When these photophores are glowing, the fish must appear much more bulky.

It is not always possible to attribute functions to the bioluminescence of animals. It is especially difficult to decide why sea pens, brittle stars, starfish and sea cucumbers luminesce. The arms of some starfish are picked out with patterns of photophores. The starfish themselves seem unable to perceive the patterns and they might attract predators rather than deter them. Similarly, some of the rat-tail fishes have a single bacteria-containing light organ in the middle of their bellies.

The production of light is a good example of convergent evolution. For an animal to live successfully in an environment requires the solving of certain problems, including how to feed, how to escape predation, how to find a mate and how to breed. In any environment there may be only a limited number of solutions to these problems. Light production is found in so many different groups of marine organisms that it must have been evolved independently many times over. Yet the chemical system producing the light is similar in the majority of organisms investigated. The structural design of lenses, shutters and diaphragms also show marked parallels.

PAGE 66
The incredible elaboration of the lures of some deep-sea fishes is well illustrated by this 15 cm (6 inch) long specimen of *Eustomias filifera*. This fish feeds on crustaceans and other fish at depths of about 800 m (2600 feet) where the number of prey species is small. The fish uses its luminous lure to tempt prey towards it, both saving energy and increasing the probability of successful capture. Such lures often resemble the food the prey feeds on, and here is one that may mimic a copepod when it is held up in front of the fish's mouth. Any disturbance near the lure unleashes a frenzy of thrashing and snapping by the fish.

PAGE 67
Astronesthes gemmifera lives at a depth of about 700 m (2300 feet). It has many features typical of deep-living fishes. The mouth is large and the teeth long and recurved. The eyes are well-developed. The lure on the lower jaw is luminous. There are ventral light organs and a small cheek photophore. The dull bronze colour is a camouflage absorbing blue-green light, which is both the colour of most bioluminescence and of the daylight that penetrates to such depths. Many *Astronesthes* species produce a bright band of luminescence along their flanks, which is possibly used as a courtship display.

LEFT
Many species produce clouds of luminescence as a reaction to danger. The little oval animals – some of which are carrying eggs in their brood pouches – are ostracods known as *Cypridina* sp. Much of the research into the chemistry of bioluminescence has been done on this type of animal. Seven glands open on the upper lip, and each gland either releases luciferin or luciferinase. Luciferin is the chemical compound that produces the light if luciferinase is present to excite it. This type of chemical reaction is common to all bioluminescence. Some midwater scarlet prawns and the squid *Histioteuthis* also produce clouds of luminescence.

ABOVE RIGHT
This close-up of the head of a fish *Pachystomias* shows the three light organs below the eye. The colour of the light reflected by each light organ from the flashlight used to photograph the animal indicates the likely colour of their luminescence. Experiments have confirmed that the most posterior light organ produces a brilliant blue-green light, while the other two organs produce red light. These cheek photophores are probably used as headlights when the fish hunts down its prey. The red photophores will illuminate red prawns, and the fish is exceptional amongst deep-living organisms in being able to perceive red light.

RIGHT
A squid of the species of *Histioteuthis* (*see next page*).

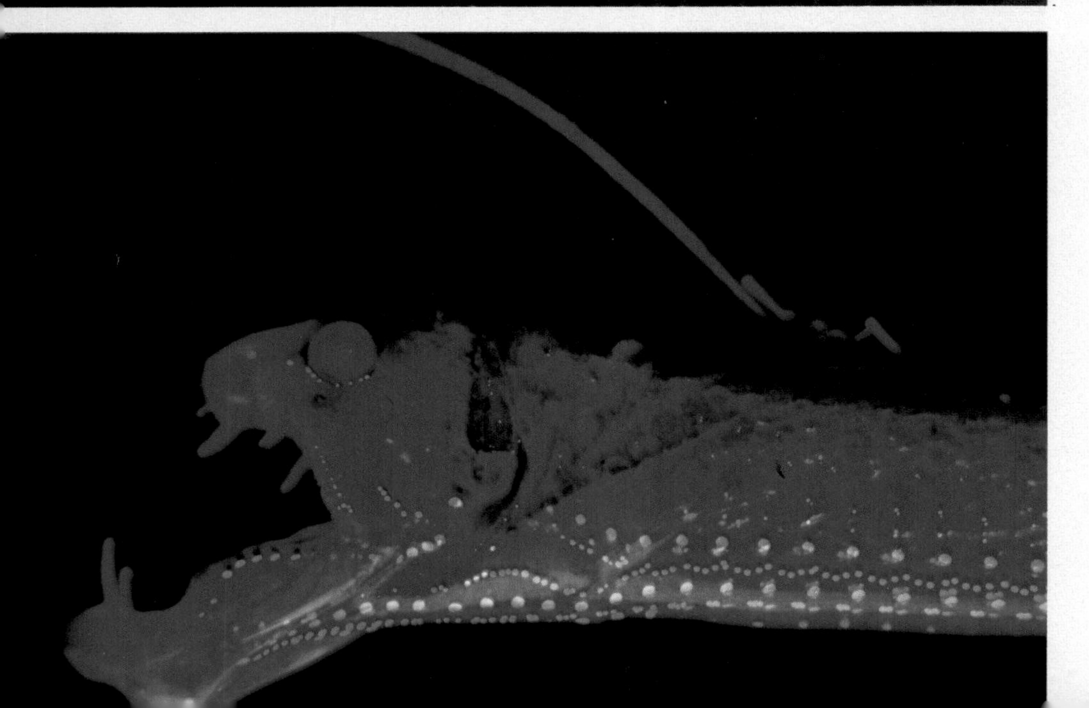

PREVIOUS PAGE
This squid is a species of *Histioteuthis* that is caught at daytime depths of 400–500 m (1300–1625 feet). The whole of the underside of the body, head and arms are studded with light organs. The squid is usually orientated with its head pointing down at an angle of 45°, and when it is in this position the photophores point straight down. The function of these photophores is to break up the squid's silhouette, making it much harder to see. Similar ventral photophores occur in many of the other animals living at these daytime depths.

LEFT
Myctophum punctatum is a lantern fish that occurs by day at depths of about 500 m (1600 feet) and migrates up into the surface at night. Along the belly is a row of silvery dots which are the ventral photophores used to break up the fish's silhouette. More light organs form a pattern over the flanks of the fish. Biologists can identify the species from this pattern, and it is thought that the fish too uses the pattern to recognize members of its own species. On the snout of the fish is a headlamp photophore.

CENTRE
The viperfish *Chauliodus sloanei* is a beautiful example of a silvery-sided fish, and lives by day at depths of 300–400 m (1000–1300 feet). The eye is large and its vision is acute. The jaw can dislocate so that the fish can swallow prey as large as itself, the large recurving teeth preventing any hope of the prey escaping. The back is darkly coloured and the belly is lined with light organs. The first fin ray is elongated and is used as a lure. The whole body is covered with a transparent jelly.

BELOW
The distribution of light organs can be studied by photographing their red fluorescence when they are subjected to ultraviolet light. Here the same viperfish is shown with the photophores fluorescing. The ventral light organs stand out most clearly. Others occur round the underside of the eye and line the inside of the mouth. The mouth photophores are believed to be an additional luring system. Minute light organs are scattered over the general body surface embedded in the jelly. These may be used in sexual display, but they may also help to make the fish look bulkier and so discouraging to potential predators.

RIGHT
Notiluca is a dinoflagellate that often occurs in big swarms during the summer months. It is frequently the source of the bright flashes of luminescence that are commonly seen in breaking waves on the shore or in the wakes of boats. The function of the luminescence is not known, but one suggestion is that any large hungry zooplankton approaching will be lit up and so be seen and eaten by predators before the *Noctiluca* gets eaten. The dinoflagellate is also a carnivore and eats small microzooplankton. Since it is only 1 mm across it cannot eat anything very big.

CURRENTS AT THE SURFACE OF
THE WORLD OCEAN IN NORTHERN WINTER

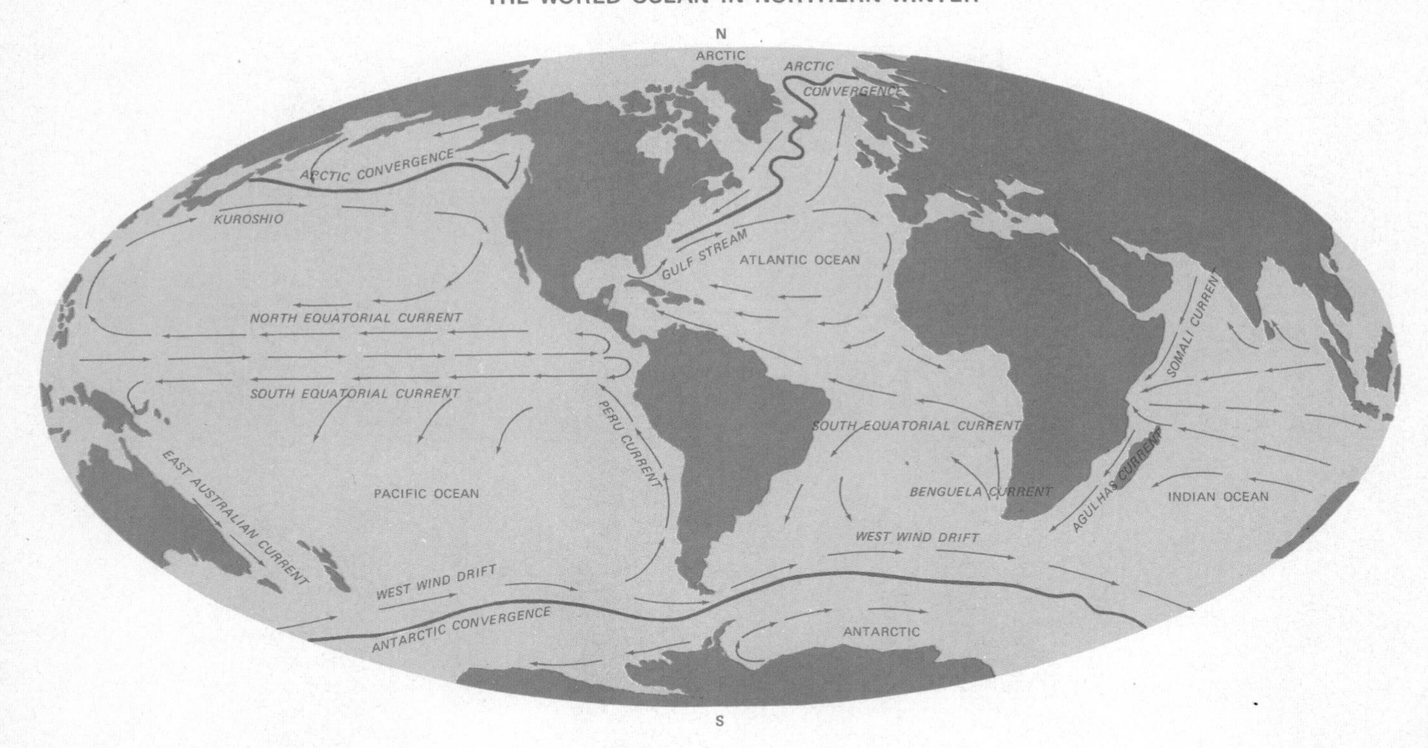

Acknowledgments

The publishers would like to thank the following individuals and organizations for their kind permission to reproduce the pictures in this book:

Afsen front endpapers, 43 top, 54 top
Heather Angel 16, 18, 22, 23 bottom, 36 bottom, 51 top, 55 bottom, 61 top
Martin Angel 14
Ardea Photographics 38 bottom (Valerie Taylor) 55 centre (Collet)
Anthony Bannister *NHPA* 54 bottom
Bruce Barnetson *NHPA* 48, 52 bottom, 64 top
Jean Marie Bassot, *Jacana* 70 centre and bottom
Sidney G. Brown 35 top
Ben Cropp, *Photo Aquatics* 1, 35 bottom, 36 top, 54–55,
Ben Cropp 38 top, 53 top right, 55 top, 65 top
P M David, *Seaphot* 12 bottom, 12 top right, 13, 14 bottom right, 15 top, 17, 20, 21, 22 right, 23 top, 24–31, 46 top, 47, 59, 60, 61 bottom, 63, 66–70 top
Walter Dawn 10 bottom
Walter Deas, *Seaphot* 19
Herman Gruhl, *Photo Aquatics* 15 bottom, 41, 52 centre, 53 top left.
W Hackmann *Photo Aquatics* 62 bottom
Peter Hill 49, 51 bottom, 65 bottom
R Lubbock *Photo Aquatics* 58
Kaven Lukas, National Audubon Society 62 top
John Lythgoe, *Seaphot* 50, 52 top left.
P Morris 11
NASA 5, 52 top right
Popperfoto 39
W Schraml, *Jacana* 32
Robert Schroeder 42, 53 bottom, 64 bottom, back endpapers
Alan and Eve Southward 40
Spectrum Colour Library 34
James Tallon *NHPA* 33
D P Wilson 6–10, 12 top left, 14 top right, 71
Woods Hole Oceanographic institution 43 bottom–45, 46 bottom
ZEFA 4 (E Grob), 37(H Lindner), 56–7 (J Grossaner)
Jacket photographs: Peter David, Herman Gruhl, Robert Schroeder, Carl Purcell.

First published in 1974 by
Octopus Books Limited
59 Grosvenor Street, London W1

ISBN 0 7064 0400 9

Distributed in USA by
Crescent Books
a division of Crown Publishers, Inc.,
419 Park Avenue South,
New York, N.Y. 10016

Distributed in Australia by
Rigby Limited
30 North Terrace, Kent Town
Adelaide, South Australia 5067

Produced by Mandarin Publishers Limited
14 Westlands Road, Quarry Bay,
Hong Kong

Printed in Hong Kong

FRONT ENDPAPERS
A school of *Haemulon*.

TITLE PAGE
Two butterfly fishes photographed off the Australian coast.

BACK ENDPAPERS
A school of coral fish.